INDUSTRIAL ARCHAEOLOGY
—— OF THE ——
PLYM VALLEY

Ernie Hoblyn

AMBERLEY

*Dedicated to Janet and John Stitson,
without whose help it would never have been written.*

First published 2013

Amberley Publishing
The Hill, Stroud
Gloucestershire, GL5 4EP

www.amberley-books.com

Copyright © Ernie Hoblyn 2013

The right of Ernie Hoblyn to be identified as the Author
of this work has been asserted in accordance with the
Copyrights, Designs and Patents Act 1988.

All rights reserved. No part of this book may be reprinted
or reproduced or utilised in any form or by any electronic,
mechanical or other means, now known or hereafter invented,
including photocopying and recording, or in any information
storage or retrieval system, without the permission in writing
from the Publishers.

British Library Cataloguing in Publication Data.
A catalogue record for this book is available from the British Library.

ISBN 978 1 4456 0554 8

Typeset in 10pt on 12pt Sabon.
Typesetting and Origination by Amberley Publishing.
Printed in the UK.

Contents

Acknowledgements	4
Introduction	6
The River Plym	8
Clearbrook to Goodameavy – Tin and Copper Mines	14
Railways	26
Dewerstone Quarries	53
Shaugh Iron Mines	70
Shaugh Lake Dry	84
Shaugh Mill	97
Riverford	103
Plymbridge Slate Quarries	113
Plymbridge and Boringdon Silver and Lead Mines	140
Marsh Mill	149
Epilogue	158

ACKNOWLEDGEMENTS

The basis for my research was an excellent archive which David Muir put together for the National Trust in the mid- to late 1980s, mainly regarding the industries and people around Plymbridge and Shaugh Bridge. This was a treasure trove, providing answers to many of my questions about the industries, and his interviews gave voices to people who had lived in the area and, in some cases, had actually been involved in some of the industries in the early to mid-twentieth century. Sadly, many of these people were no longer in a position to tell their own story by the time I became interested. My grateful thanks go to the Trust for allowing me to quote from his interviews and include some of his photographs.

I love maps, and the First and Second Edition 25-inch Ordnance Survey maps proved enormously useful in my research. Plymouth Central Library has copies of all the local early OS maps and also all the local newspapers, dating back to the eighteenth century. 'Local' in this context is a relative term; in the eighteenth and early nineteenth centuries, the *Sherborne and Yeovil Mercury* and the *Exeter Flying Post*, among others, carried adverts regarding local industries.

Ancestry.co.uk was also enormously helpful, allowing me to give names and backgrounds to the often forgotten, faceless people I was writing about.

Plymouth and West Devon Record Office is just down the road from me. With the aid of the immensely helpful staff there, especially Deborah Watson and Alan Barclay, I discovered many answers to questions which had otherwise proved elusive.

The Dartmoor Tinworking Research Group were enormously helpful with mining research, and not just about tin mining. My special thanks to Stephen Holley for his advice and for reading and correcting the draft of the Clearbrook chapter.

The China Clay History Society gave me a great deal of assistance regarding the china clay dry at Shaugh Bridge and allowed me to use some of their photographs. My special thanks go to John Tonkin from St Austell who took

Acknowledgements

the trouble to talk me through the processes and working methods, as well as giving me access to his own research.

John Boulden, John Bowler and John Luscombe for their invaluable help and photographs.

Brian Moseley and his excellent Plymouth Data website provided many answers to my questions and suggested other areas of research.

Darren Payne for giving me access to his history of the Gullett family.

Antics Ltd for allowing me to use a picture of their OO gauge model of a Hoare Brothers railway wagon, produced by Dapol for Antics model shop.

Science & Society Picture Library for allowing me to use the lovely picture of Brunel's original Cann Viaduct.

Bibliography

A. K. Hamilton Jenkin, *Mines of Devon*.
H. G. Dines, *The Metalliferous Mining Region Of South-West England*.
C. F. Barclay, *Mines of the Tamar and Tavy*, a collection of his mine reports.
Raymond Burnley, Roger Burt and Peter Waite, *The Devon and Somerset Mines (Mineral Statistics Of The United Kingdom, 1845-1913)*
Maurice Dart, *Narrow Gauge Branch Lines – Devon Narrow Gauge*.
Mike Brown, *Dartmoor Field Guides*.
John R. Smith and the Royal Commission on the Historical Monuments of England, *Shaugh Bridge China Clay Works, An Archaeological Evaluation For Dartmoor National Park Authority*.
Exeter Working Papers in Book History, Shaugh Mills.
Various Directories of Devonshire.
Darren Payne, and his history of the Gullett family.
Eric Hemery, *Walking the Dartmoor Railroads*.
Bernard Hill, *The Branch*.
John Binding, *Brunel's Cornish Viaducts*.
Brian Lewis, *Brunel's Timber Bridges and Viaducts*.
Bert Shorten, *Plympton's Old Metal Mines*.

INTRODUCTION

As this is an Introduction, first let me introduce myself. I was born in 1951 and brought up just across the River Tamar, in south-east Cornwall. Unusually for that era, for a time my parents and I lived in a cottage on the Mount Edgcumbe estate which was at least a mile from anywhere and thus at that time had neither electricity nor mains water, it being deemed far too expensive to supply either over such a distance for just one cottage. The cottage was, however, surrounded by woodland and wildlife, including deer, and I loved living there. As well as a life-long love of nature, that upbringing gave me an insight into how hard life was in earlier times, lacking the amenities we all take for granted today, an insight which few of my age might have.

My background is in engineering which possibly explains a lot. I can't look at anything without the questions starting; 'How did they build that? How did they get the materials here? Why did they build it here? What is it for?' Consequently, having been walking around the Plym Valley for a while and seeing various ruins of buildings in strange places, I wanted answers to all the above questions. It took me a long time, but eventually I started to find some answers. Those answers just made me more curious so I kept on digging until eventually I was satisfied that, while I might not know everything, I knew something! Then I started thinking that other people might want to know the answers to these same questions as well, so I started writing. Many years ago I wrote articles for magazines, so I had some idea how to write, but this was on a totally different scale. I was used to writing 2,500 words for an article; here, I was faced with an open-ended commitment. Some of the chapters just grew and grew! The more I wrote, the more I realised how little I knew and the further I had to dig to find answers before I could write some more.

Most of the industries in the Plym Valley existed during the period from the mid-eighteenth century to the late nineteenth or early twentieth century, roughly from the start of the Industrial Revolution to the late Victorian era. Few survived very far into the twentieth century but luckily, thanks to a project undertaken by the National Trust in the 1980s to interview the people who had lived in the valley, I had access to the memories of a few people who remembered some parts of them. Today, little apart

Introduction

from the ruins of buildings survives as witness to all the hard work done by many people over many years. My aim is to bring a few of these ruins to life and tell their story.

I have tried to stick to my brief, writing about the industry in the valley which runs north from the Laira estuary to Clearbrook. In fact, this valley contains in its length both the Plym and Meavy rivers, but to keep things simple I will refer to it as the Plym Valley because the Meavy is a tributary of the Plym. My definition of 'in the valley' was purely arbitrary; I have included some sites which are some way from the river because I felt they had a definite connection, and excluded others, particularly the mines on Roborough Down, because I felt they would have led me further and further afield. I have largely ignored the enormous clay works near Cadover Bridge and the ancient tin streaming which took place mainly on the high moor; others with far more knowledge have written at length about these.

With this book I am aiming to provide the answers I was looking for when I came across remains of buildings and structures many years ago, and to appeal to the curious who, like me, want to know why, when and how? My intention is only to give an overview of the industries. Of necessity, I have had to go into a certain amount of detail in order to explain why buildings were there, what they did and how they worked. I have tried not to go into so much detail that the reader becomes bored in a few pages; those whose interest has been awakened and wish to know every tiny detail will find all they need in specialist publications or by joining specialist groups. Some of these groups are mentioned in the Acknowledgements.

Finally, the dimensions are given in miles, yards and feet. This is for two reasons: firstly, because although I can work in metric measurements I can't think in them, and secondly because that was the way things were measured at the time. Dartmoor Gauge was 4 feet 6 inches, not 1.371 metres; the distance between fixing chairs for fish-belly rails was 3 feet, not 0.914 metres. I'm not about to rewrite history just to comply with EU regulations!

This book has involved a huge amount of research and I have received an enormous amount of help from a great many people. I hope I have covered all of them in my Acknowledgements; if I have missed anyone then please accept my sincere apologies. I have also done everything within my power to obtain the correct permissions from the appropriate people for the use of all the photos. If you own the copyright to any photo and I have credited it to someone else, please accept my apologies, let me know and I will do my best to correct this.

If, having read this, you feel you want to blame someone for the creation of this book, blame my dogs! If I hadn't had dogs I probably would never have spent so much time walking them in the Plym Valley, then none of this would have happened.

THE RIVER PLYM

The city of Plymouth as we know it today exists on an area of land between two rivers – the Plym, which flows down the eastern side of the city, and the Tamar, which flows down the west side. Given that the Tamar is not only the larger river by far, but also forms the border between Devon and Cornwall, it appears perverse to modern eyes that the city should be named after the smaller, less well-known river.

In fact, the settlement which grew from a small fishing village to become the large city we now know as Plymouth was located around the area now known as the Barbican or Sutton Harbour, which is a sheltered harbour at the *mouth* of the River *Plym* – hence Plymouth.

This river flows down from its twin sources high on Dartmoor, along the eastern and then the southern sides of the city. I fear that few people in the city today think of, or even in some cases know of, the existence of this river; it is overshadowed by the larger and more famous Tamar. If they know of the Plym at all, they may know of its tidal estuary, the Laira, which stretches along beside the Embankment, one of the main access roads into the city. Those who enjoy walks in the country might have visited Plymbridge or Shaugh Bridge, two of the best-known access points in the Plym valley from which to enjoy the local area.

Few know of the history of the valley of the Plym and just how much industry existed there in the past.

The River Plym has its source high on Dartmoor at Plym Head, south of Fox Tor, within a mile of the source of the River Erme and some 12 miles as the crow flies from Plymouth Hoe. Where the Erme heads south towards Ivybridge and the sea, the Plym winds its leisurely way south-west past the warrens at Ditsworthy and Trowlesworthy, skirting the edge of the Lee Moor china clay works and on to Cadover Bridge. From here the river rushes down through a narrow, steep-sided, rock-strewn valley to Shaugh Bridge, where it meets the River Meavy.

The River Meavy (originally Mewi) starts its life further north, in fact just south of Princetown and about 14 miles from Plymouth Hoe. From its source, it flows among the bleak moorland valleys until it reaches the now flooded valley

The River Plym

which is Burrator Reservoir. The Meavy is the main source of water for this reservoir, and also for the two leats, Drake's and Devonport, which preceded that reservoir in supplying the people of Plymouth with their water.

The overflow water from the reservoir then continues on its south-westerly course down the old Meavy valley, past Meavy village, and turns south below Yelverton. From Clearbrook there is a broad, lazy stretch down past Goodameavy, but below there the Meavy also rushes through a narrow, boulder-strewn area before reaching the confluence with the Plym at Shaugh Bridge.

The Meavy is the longer river and arguably, even despite the amount which is abstracted at Burrator, still brings more water; why is the river system named after what seems to be the junior river? The only clue I have seen is in eighteenth- and nineteenth-century records and maps where the higher section of the River Plym around Cadover Bridge is frequently referred to as the Cad rather than Plym, hence the name Cadover; maybe in the distant past the naming was different. Maybe back then the rivers Cad and Meavy joined at Shaugh Bridge to become the Plym. Whatever the reason, it was well in the past and is beyond the scope of this book.

From Shaugh Bridge the now unified river heads more or less due south through Bickleigh Vale, past Riverford and Plymbridge, passing two weirs on the way. Below Plymbridge, wider and slower now, the river meanders across flat marshy ground, which is within the limit of tidal flow and hence the river was prone to flood, until it joins the Laira at Marsh Mills. The Laira, depending on the tide, is either a beautiful tree-lined expanse of sparkling water between the eponymous area of Plymouth and the lovely Saltram estate opposite, or it is a vast mud flat providing food for thousands of wading birds. Either way, on a bright sunny day it can be a grand view. Rounding the last bend between Oreston and Cattedown, it flows past Mount Batten pier and into the majestic spectacle which is Plymouth Sound, where it joins the River Tamar.

So now that I have described the river, where was the industry? Apart from the Lee Moor china clay works, still very productive after almost 180 years of operation, there is no industry in the valley today, with the possible exception of some forestry work carried out by the Forestry Commission. The upper reaches on the windy and rain-swept moorland seem unlikely industrial sites; surely only hard-working farmers and warreners could scratch a living up here. Well, even in ancient times there was tin streaming near Trowlesworthy, and piles of stone waste are still there to be seen. In more modern times, the huge Lee Moor china clay works show what can be done up here despite the inhospitable environment. Lee Moor is outside of my area of interest, which is the valley from Clearbrook down to the estuary. I will not be writing directly about Lee Moor or the tin streaming, which have both been well documented by others, but both have connections which will appear later.

As Lee Moor shows, however, where minerals are found, some hard-working people will do their best to extract and exploit them no matter how difficult that might seem. Around Clearbrook tin and copper were found, minerals more usually connected with the Tavistock area, and several mines were dug in the area. Between Goodameavy and Bickleigh the river flows through an area of granite, an extremely hard and tough rock much prized by builders. Around Shaugh Bridge there was a clay drying kiln, an offshoot of the Lee Moor clay works, which was still operating until the mid-twentieth century. Also in that area a sizeable deposit of iron ore was found, not something with which we would usually associate this area, but nevertheless several mines were in operation there, along with a brick works which used the ore. Shaugh Mill, below Shaugh Bridge, had been a paper mill until the late eighteenth century and then became a grist mill. Further down the valley between Riverford and Plymbridge, slate abounded along with blue elvan, a hard-wearing stone used in road building. Several quarries were in operation from the seventeenth century; some were still in existence into the twentieth century.

Also in this area was another type of mining lead, zinc and even silver were to be found in quite large quantities, sufficient to keep these mines running for many years during the mid-nineteenth century.

Below Plymbridge there was little in the way of industry, but there was the Marsh Mill, which gave its name to that part of Plymouth. Large and growing conurbations such as Plymouth and Plympton would have needed bread and bread meant an increasing demand for mills, of which there was a great number in the eighteenth and nineteenth centuries. The Marsh Mill was one of the largest, most successful and longest-lived of them. Fed with water via a leat from the weir at Cann, it was open in 1582 and still in use into the twentieth century, eventually closing in 1929.

Finding and extracting these minerals in such inhospitable areas is all very well, but what to do with them once extracted? Plymouth has been a famous port with worldwide links for centuries, but the port of Plymouth is a long way from the Plym valley, with many hills and marshes between. The presence of useful minerals calls for a transport system to carry them to the port, and other hard-working people were willing to expend money and energy building these transport systems. The roads in this area were (and still are in many cases!) very poor, so some kind of railway was needed, and soon there were railways in abundance. First, in 1823, came Sir Thomas Tyrwhitt's Plymouth & Dartmoor Railway, which despite its grand-sounding name was actually little more than a horse-drawn tramway. It was originally built to service his quarries near Foggintor and return with fertiliser to enrich the poor Dartmoor soil, but was also available to carry materials for others along the way. A later branch from this line was built to carry slate from Cann quarry down to Laira and on to Sutton Harbour. Another tramway, including a long inclined-plane section, serviced Lee Moor, bringing the china clay down to join the Plymouth & Dartmoor Railway at Crabtree.

Later, in 1859, the Tavistock & South Devon Railway opened. This was a fully-fledged railway as we would know it today, carrying goods and passengers from Plymouth at first as far as Tavistock, but by 1865 it was extended to Lydford and Launceston. This would have made a huge difference to the Plym valley as a whole, not just because of the increased access it allowed, but also because this railway was built along the line of the valley, staying close to the river from Marsh Mills all the way to Clearbrook. The embankments, cuttings and viaducts were a massive undertaking and are still there today for us to see.

What of the people who lived and worked in these remote places? Life for those who worked in these industries was harsh. According to a study published by the Cambridge Group for the History of Population, the life expectancy of people in 1840 was the early to mid-forties, as opposed to our current life expectancy which has risen steadily over the years until it is now around eighty years. While there are many reasons for this, not least of which are the medical improvements which have controlled previously fatal diseases, the fact that their life was so harsh must have been a major factor – they were simply worn out by the time they had reached what we would call middle age.

The men who worked in mining or quarrying could expect to have to walk to work, often several miles, whatever the weather. Feeling cold and frequently being wet was just a part of life, even at home, where it would often have been too expensive to heat more than one room – if indeed they lived in more than one room. They would be expected to do very hard physical work, in some cases for up to 12 hours, then walk home. All these industries were labour intensive – at that time there were no machines to do much of the work. They would be likely to spend their days swinging a sledge hammer which could weigh up to 20 lbs, drilling holes in solid rock with a long chisel. These holes would then be packed with explosive, gunpowder in the early days and dynamite later, to remove large chunks of rock, either for use as it was or to be refined for the ore it contained. This rock would then have to be moved, which obviously involved more back-breaking work. Twenty years of that kind of work would turn a fit fifteen-year-old into an old man. There was no alternative, though: with no unemployment benefits or pension, you either worked or you starved, so people often worked until they died.

For women, the work may not have been so heavy but it was nonetheless relentless and hard, with none of the modern conveniences which we take for granted. Staying warm meant building a fire in the early morning, probably using wood which the woman had cut and chopped herself. Having a drink meant getting water from the well. Eating meant walking to the nearest shops, which could easily be many miles away, or growing food in the garden and even keeping rabbits, chickens or a pig. Washing clothes was a day's work, given the time it took to fetch and heat the water, then wash everything by hand, put it through a mangle to get out as much water as possible then carry it outside to dry, and of course with the lack of heating they wore many more clothes than we would today, although possibly they changed them less frequently!

Unusually for someone who was born in the 1950s, I have some insight into just how hard their lives were because I lived for five years in a cottage on the Mount Edgcumbe estate in Cornwall which at that time had neither electricity nor running water. The nearest village, Cremyll, was a mile away through woods, hills and valleys and the cost of running electricity and water so far for one cottage was prohibitive. We had to pump water from a well using a hand pump. We used paraffin lamps for lighting and an open fire for heat. Having a bath meant pumping the water into the copper, which was in an outside washhouse, then lighting a fire under it. When the water was hot, which could take an hour or two, we could run it into a tin bath and my parents and I would take turns in the hot water. This was not something to be undertaken lightly so a bath was a weekly affair! As I said, Cremyll was a mile away, but at least there was public transport available there to get to school or the shops, something which would not have been the case in the 1850s.

By the early twentieth century most of the industry in the Plym Valley had closed but the railway meant that the people of Plymouth, who by now had more leisure time, had easy access to a beautiful natural area. The so-called Woolworth's Specials ran at weekends, allowing Plymothians to explore the whole area from Marsh Mills for the princely sum of *6d* (2½p). Many places along the valley catered for these weekend visitors, with tea rooms and kiosks operating right up to the Second World War.

So, there is much to see in the area. Come with me and explore.

The book has been split up into sections referring to the various areas, the idea being that readers can find out what went on in a particular area they know well or intend to visit. Anyone who reads it from start to finish will find a degree of repetition because some things apply to more than one area; for this I apologise, but I thought it better to put something in too often than have it missed because it was only mentioned in another chapter.

A map showing Plymouth and the Plym Valley.

CLEARBROOK TO GOODAMEAVY – TIN AND COPPER MINES

My story of the industry in this area begins where the River Meavy, after first meandering across open moorland and then feeding the enormous Burrator Reservoir, turns south near Clearbrook and begins its journey down through the valley to the sea.

For many years I have enjoyed walking around the area of moorland above Clearbrook called Roborough Down, and the many trenches, pits, ridges and mounds which can be found all over the area always intrigued me. Who or what caused them? Why were they there? Now I know. They were caused by the local industry – mining.

Most people tend to think of copper and, more particularly, tin mining as being a Cornish industry. Those with knowledge of local history will know it was also something which took place in the Tamar valley, particularly in the area around Tavistock, Calstock and Gunnislake. Indeed, during the nineteenth century the Devon Great Consols mine, on land belonging to the Duke of Bedford between Tavistock and Morwellham, was one of the largest copper mines in the world. Morwellham Quay was developed during the nineteenth century to allow the export of its enormous output, plus that of other mines in the area, but what does that have to do with the Plym valley? Well, surprisingly, both types of mining also went on here.

Tin has been recovered from Devon and Cornwall since ancient times. There are records of the Phoenicians voyaging to the south-west of England to buy tin 200 years before the birth of Christ. Joseph of Arimathea, supposedly Christ's great-uncle, was reputed to have been a tin trader who journeyed here, and some even suggest that he brought Christ with him. Early tin-smelting ovens, in use since the second century BC, are known in Cornwall as 'Jews' Houses'.

As I mentioned in the previous chapter, there was some tin streaming in ancient times, near Trowlesworthy on the Plym for instance, and at Black Tor Falls on the banks of the Meavy. In olden days this tin streaming would have been surface working, with men digging up the ore where it appeared on or near the surface and using water to sift out the heavier tin ore from the lighter sand and soil. When

Clearbrook to Goodameavy – Tin and Copper Mines

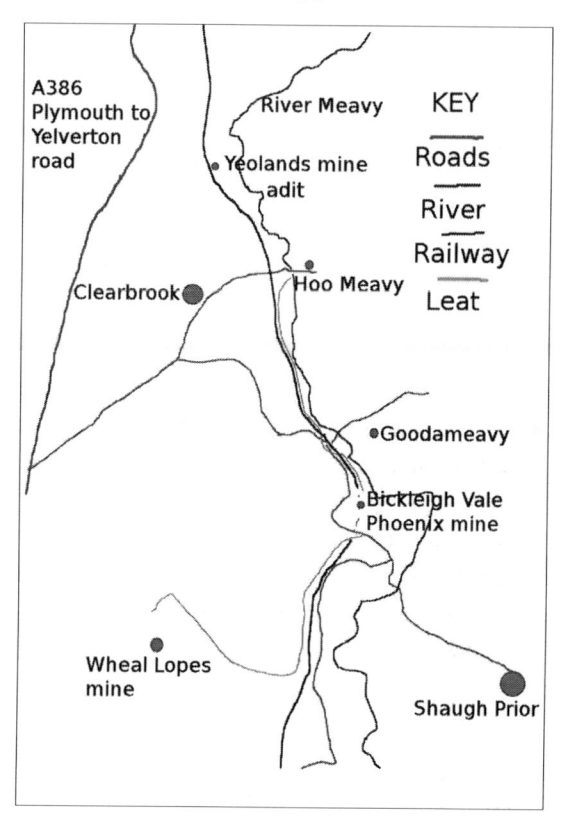

A map showing the area around Clearbrook and Bickleigh.

we look at the amount of soil which was moved in these operations, it obviously represented a huge amount of effort, with long hours of back-breaking work in the harsh climate of the high moor, but it was not really mining as we think of it today.

There is documentary evidence that large quantities of tin have been recovered from south-west Dartmoor since at least the twelfth century, and the two centuries following that have been described as being the boom years of tin production. This is because the ore the early miners were recovering at or near the surface tended to be of very high purity. Paradoxically, the deeper they needed to go to find the ore, the less pure it was and more refining was necessary to recover the tin.

There were tin lodes (as the accumulations of ore are called) beneath Roborough Down, the moorland south-west of Yelverton, and they had been worked since before Elizabethan times. To the east of Hoo Meavy bridge, in Hoo Meavy itself, there is a row of tin miners' cottages which are claimed to date from the sixteenth century. A. K. Hamilton Jenkin, in his book *Mines of Devon*, says there were ten lodes running east to west across the moor between the main Yelverton to Plymouth road, now the A386, and the River Meavy. These were mainly between Yeolands Farm, which is to the west of the road, and the present site of the village of Clearbrook. He quotes a letter from Sir Walter Raleigh to Sir Robert Cecil, then Secretary of State to Queen Elizabeth I, dated 15 November 1600, in which he says: 'A gentleman, Mr Crymes, hath erected certain clash mills on Roburghe down to work the tynne which in that place is got with extreme labour and charge out the ground.' (The Crymes family were the local landowners at the time.)

At that time they were probably still recovering the tin ore which was at or near the surface, so it was largely just dug out of the ground, although Dr Tom Greeves, who was awarded his doctorate for his study of the Devon tin industry, says that vertical shafts with horizontal drainage adits were 'probably commonplace on moorland Dartmoor by the sixteenth century'. If the miners of 1600 thought the tin was 'got with extreme labour and charge out the ground' at that time, they didn't know how lucky they were! Deeper mining would have been difficult or impossible then without some means of pumping out the water, of which Dartmoor has little shortage.

Yeolands Mine

The lodes at or near the surface would have been reached relatively easily and had been pretty well worked out by the beginning of the nineteenth century. What is for sure, though, is that although the surface deposits had been largely worked out, there was still tin to be found beneath the ground of Roborough Down. Clearbrook did not exist as a village until around the middle of the nineteenth century – prior to that, there was little there but three farms – but by the 1850s tin mining was extensively carried out in the vicinity and according to the 1851 census the majority of Clearbrook's adult male inhabitants were tin miners. In 1848, according to Charles Thomas, manager of Dolcoath Mine, who wrote a report on the mines in this area, between Yeolands Farm and the River Meavy at least three shafts – named Engine, Edwards and Odgers shafts – were in production, recovering tin from three separate lodes. Yeolands Farm is on Yeolands Lane, off Golf Links Road, to the west of the A386, on the way down to Milton Combe and Lopwell.

There is more than a little confusion over the names of this mine or mines. At various times they were referred to as Yeolands Mine, East Yeolands, South Yeolands and even Plymouth Wheal Yeolands. H. G. Dines, in his book *Metalliferous Mining Region of the South West*, refers to only the one mine, which he calls Yeolands Consols, and includes the other names as parts of that

Clearbrook to Goodameavy – Tin and Copper Mines

mine. In fact Yeolands Consols started in January 1851 as an amalgamation of the existing mine companies; the new company set out with the intention of pursuing their operations 'vigorously'.

Hamilton Jenkin says that the engine shaft of Yeolands mine, which was 500 yards east of Yeolands Farm (roughly where the A386 passes the farm today), was said in 1853 to be 64 fathoms (384 feet) deep, with four levels off it in an east–west direction, all of which were found to be productive. Dines, in his book, describes three lodes running east to west, known as North, Main and South lodes. Of these, Main Lode appears to have been the most productive, running equally some 380 fathoms (2,280 feet) both west and east from Engine Shaft. The end of this lode is visible at the surface because it is the portal of its Deep Level Adit (an adit is a drainage tunnel), which is between the old Great Western Railway line and the River Meavy. C. F. Barclay, who surveyed the area in the 1930s, says there is believed to be an old shaft on the eastern side of the River Meavy and some open workings in the wood, near to Olderwood Farm, but he does not say anything further on the subject and the maps show no evidence of these works, which might have been simply trial digs.

Main Lode is described by Hamilton Jenkin saying:

> Being of great size, 8 to 10 fathoms (48 to 60 feet) wide in places. Between 1851 and 1856 some £13,000 worth of black tin had been sold and two steam engines were employed, one 22 inch engine for hoisting from the shaft and the other, a larger 36 inch engine for pumping and operating the 24 stamping machines needed to process the ore.

One of these engines mentioned above had been operated by the previous company, but a new one was installed as part of Yeoland Consols' 'vigorous' operations and they increased the number of stamps worked by the old engine to thirty-six, with another twelve worked by the new. They also increased the amount of underground working and according to reports which regularly appeared in the *Plymouth & Devonport Weekly Journal*, by the end of 1851 all was ready so that, 'The adventurers in these mines will ere long reap the fruits of the untiring perseverance with which operations have been carried on, and a steady and increasing monthly return will from this time forth establish the character of this speculation on a firm and substantial basis.'

Devon and Somerset Mines (Mineral Statistics of the United Kingdom, 1845–1913) by Raymond Burnley and Peter Waite gives even more, if slightly contradictory, information about the output of the mine. They give a total of 292 tons of ore removed between 1852 and 1857, producing £15,730 income.

The history of tin mining has always been dependent on the price of tin. Given the huge amount of effort needed to recover and refine the ore, it was only worthwhile if the price was good. This explains the sporadic nature of the working of the mines. In spite of the relatively high returns in the preceding years, in March 1857 the price of tin dropped, the mines were abandoned and the sett and machinery put up for sale. The difficult conditions underground, mentioned in various *Plymouth & Devonport Weekly Journal* reports, plus the cost of having to use two steam engines all day long for hauling and stamping, cannot have helped the economics of these mines.

Correspondingly, among the inhabitants of Clearbrook, the number of miners rose and fell with the value of the various ores. According to the census of 1851, tin miners made up the majority in the village; there were twenty-five adult males living in the fifteen houses, and of them fourteen were tin miners. In 1861 there were just six copper miners and one tin miner. In 1871 there were no miners at all, and in 1881 there was one copper mine labourer and also John Dester Canning, who was manager of the iron mine and brickworks at Shaugh Bridge, obviously a special case! Surprisingly, there were four tin miners shown in the 1891 census, although they were rather outnumbered by the rest of the thirty-one adult males living there!

One of Mike Brown's Dartmoor Field Guides includes a report by Captain Joseph Eddy of Bottle Hill mine near Plympton by Captain Joseph Eddy, dated 1870. Captain Eddy had been asked to give an opinion on the causes of the previous abandonment of the Wheal Yeoland sett and he firstly described geological problems which would have required a new vertical shaft to put right. Secondly, the fact that water had to be pumped out of the mine using steam power, then pumped again to provide power for the stamps, was hugely expensive at a time when the price of tin was falling. He recommended working the mine from the riverside adit, where water power might be available, rather than from the top of Roborough Down, where it was not.

By 1881 a new company with capital of £60,000 had been formed to work Yeolands Consols, the intention presumably being to work the mine as recommended by Captain Eddy, as quoted above. Very little appears to have been done apart from sinking a new shaft. By 1884 there was reported to be no power from either steam or water to operate the mine. The former manager of the mine, Captain Richard Williams, wrote a letter stating that in his opinion, 'The lode is large, the ground bad and the average yield of black tin does not exceed 4 lbs per ton. To say that the lode improves with depth is not true.'

In 1886 a new shaft was dug which went down through 'old men's workings' – in other words an area which had already been mined, so was unlikely to produce

Clearbrook to Goodameavy – Tin and Copper Mines

Yeolands House today.

The adit portal, now a water supply.

Industrial Archaeology of the Plym Valley

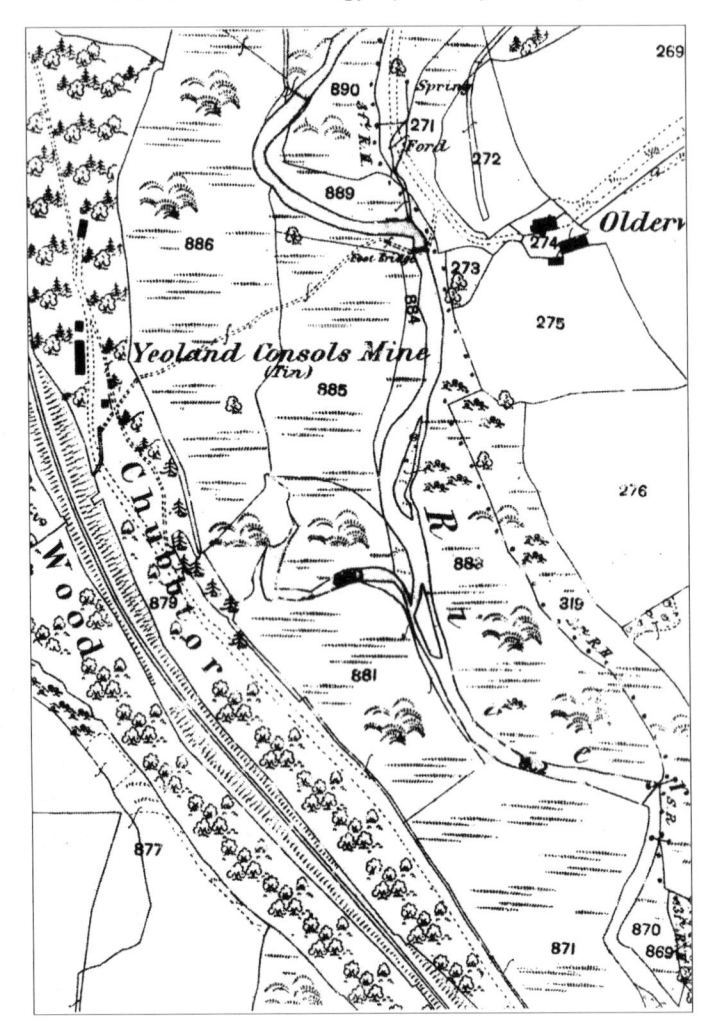

A map of the area around the Yeoland Consols mine in 1865.

Yeolands name board. (*John Bowler*)

Above left: A slate used as a tally. (*John Bowler*)

Above right: A selection of names from the slate. (*John Bowler*)

good results. This was the peak year for employment at the mine, with a total of fifty-seven men working there. The following year more capital was promised to sink a new shaft, but by the end of that year none had materialised and the company funds were more or less exhausted, although they managed to limp on for a few more years. History appears to side with Captain Williams; between 1883 and 1888, the mine produced 177.6 tons of ore, worth £10,195, considerably less than had been produced between 1852 and 1857.

By 1887 the company was in liquidation according to Burnley and Waite, although they quote output figures for one year more. According to the same source, by 1889 the company was employing just one man, presumably a caretaker; it was suspended in 1890-91 and abandoned in 1892, the owners having gone to America! C. F. Barclay, when he surveyed the mine in the 1930s, found lodes varying from 1 to 2 feet wide, but samples taken from various places gave poor results. It does appear that Captain Williams was correct in his assessment of the poor quality of the ore, but there also appears to have been gross mismanagement of the company. Sadly, this was not uncommon in the mines of the area.

As I said at the beginning of this chapter, evidence of the early surface workings abound on Roborough Down. Anyone walking on the moor, anywhere between Yelverton to the north and Roborough to the south, and between the A386 in the west and the River Meavy, will find in many places strange pits, trenches and mounds which show where the miners worked the ore near the surface many years ago. Anyone wishing to see the best surviving evidence of Yeolands Consols mine will have to walk half a mile or so along the footpath which runs parallel to the River Meavy, going north from Hoo Meavy bridge. This will take you to the portal of what was called Deep Level Adit, about 50 yards short of Chubtor Bungalow and Yeolands House, the former Mine Captain's residence. The portal is now blocked by a grating and dammed further in; the water which the adit was originally dug to drain from the mine is now the source of excellent drinking water for these two houses.

In the privately owned woodlands in front of the mine portal are the remains of a building which is thought to be a calciner, a type of furnace where the ore was heated to high temperature as part of the initial refining process, plus other buildings, one of which might be a waterwheel pit, and several circular depressions called buddles. These were troughs used to wash the ore and allow the heavier, tin-bearing ore to settle out from the lighter, less productive waste. One of the fields below this area appears to be very much higher than the surrounding ones, the reason being that this was where the spoil was dumped!

John Bowler, who currently lives at Yeolands House and who furnished me with much of the information quoted above, has found a couple of interesting artefacts relating to the mine. One is an original name board which John has dated to 1891, shortly before the mine was abandoned.

The other is a slate which had been used as a tally, with names and figures, presumably individual production figures, scratched on it. Judging by the holes and marks on this, it had also been used for roofing, although whether previously or subsequently is difficult to prove!

Bickleigh Vale Phoenix

In 1858 the Plymouth to Yelverton section of the Tavistock & South Devon Railway was being constructed along the length of the valley. Approximately three-quarters of a mile south of Clearbrook, further down the valley of the Meavy, near Goodameavy Bridge, the builders of the line had reached the solid granite obstruction of the Leebeer Ridge, half a mile from Shaugh Bridge. The only way to continue was to build a tunnel through this ridge. In the process of digging the tunnel, a lode of copper was found, running east to west across it, about 100 yards in from the northern portal below Leebeer Farm. Although found by accident, this lode was too good an opportunity to miss, so, according to Hamilton Jenkin, in 1859 a prospectus was issued to form a company to exploit it. The prospectus which described the copper lode as '12 feet wide where it had been opened at the surface', seems to be at variance with the original report on the find which, according to Hamilton Jenkin, described it as 'fully 4 feet wide' in the roof of the tunnel, reducing to '1 foot, very hard, no ore' near the floor. Also, a lode of tin 4–6 feet wide had been found near the entrance of the tunnel. As often seems to be the case in a nineteenth-century prospectus, there is far more hyperbole than would be allowed by our modern Trade Descriptions Act.

Given the proximity of the river to provide water power, the road which passes within 100 yards and of course the railway passing immediately beside it, this mine would appear to have had much to recommend it to investors. The mine was originally called the Tunnel Mine, but the name was later changed to the Bickleigh Vale Phoenix Mine. Sadly, in spite of the hyperbole in the prospectus and the advantages named above neither name enthused the investors, and little or nothing was done.

An adit, presumably the original excavation, which starts from a point some 110 yards in from the northern mouth of the tunnel and runs approximately 200 yards to the east, is still visible today, although flooded in places to a couple of feet in depth. The entrance is blocked by a grating which has been bent out of the way and people have been in there, judging by the pictures which appear on the internet. The entrance to another adit is some way further down the valley, away from the railway line in the wood. This adit, which I am told by those who have explored it is L-shaped and also flooded, presumably joins the east–west lode as it continues on the west side of the tunnel. There were some surface workings carried out to the east of the tunnel, with what looks like a shaft and spoil tip just below the leat which runs parallel to the tunnel. This seems to line up with the end of the adit and would have made a much more accessible (and safer!) entrance to the mine than that in the tunnel. Following the line of the adit further toward the river, there is a large pit, a possible surface excavation, by the riverside which is still visible today. Nothing further appears to have been done and in 1862 operations were suspended.

The surface workings are easily visible to anyone walking from Goodameavy Bridge down the west bank of the river, near the bend in the river more or less opposite to Dewerstone Cottage. There is a large, circular pit dug into the higher ground 10 yards from the bank; the spoil tip and capped shaft are on the hillside above. The entrance to the adit can be seen by walking along the cycle path through the tunnel, about 100 yards in from the northern entrance on the eastern side.

WHEAL LOPES

Approximately a mile further south from the site of Bickleigh Vale Phoenix mine was Wheal Lopes, named after the landowning family of the Maristow Estate. The head of the family was originally a baronet, but in 1938 became Baron Roborough.

This mine, with lodes of copper, zinc ore and some tin, was mentioned in 1760, when a lease for mining was granted, but nothing is known until the 1820s. By 1821 the lode had been fairly extensively explored, with one level producing 140 tons of good copper ore according to A. K. Hamilton Jenkin, who provides most of the available information.

Power to pump the mine and operate machinery was provided by a 30-foot-diameter waterwheel fed by a leat which, at enormous expense, had been brought all the way from the River Meavy at Hoo Meavy bridge, below Clearbrook. This leat, which runs parallel to the railway, met with the same problem which the railway had – the Leebeer Ridge. It also needed a tunnel in order to pass through this granite ridge, in this case one 500 feet long, emerging on the Bickleigh side of the ridge; from there, the leat crossed the railway by an aqueduct above the southern portal. Despite the enormous outlay involved in building this leat, the upkeep was much less than the cost of running steam engines, as mentioned previously.

The mine was advertised for sale in 1823 and closed in 1825. In 1840 it was acquired by the Plymouth and Dartmoor Mining Company, which installed new machinery and developed the mine to much deeper levels. The western part of the mine was most productive, producing in early 1844 most of the 552 tons of ore which the *Mining Journal* recorded had been mined in May and which made £2,666.

As always, the price of minerals fluctuated and later in 1844 the price of copper dropped from £107 to £90 per ton, making the mine uneconomical, so it was again sold up and closed. In 1856 the price increased to £130 per ton, so the mine was reopened and continued to limp along under various guises. Mismanagement seems to have been the cause of most of the final problems; in 1868 the mine was finally closed and the company wound up.

Just before this winding-up of the company, in December 1867, a case concerning the mine came before the Tamar and Plym Fishery District Board of Conservators. This was a group of gentlemen with special interest in the fish stocks of the rivers; mainly of course they were men who owned fishing rights to various sections of the rivers. The chairman, Mr Soltau Symons, rented fishing rights from the Earl of Morley and had made great efforts to improve the stock of salmon, trout, etc. in the river. In the six months prior to the meeting, thousands of fish had died in the river, as far upstream as the Wheal Lopes mine but not further. The Conservators therefore attempted to lay the blame for this poisoning on the mine owner, Sir Massey Lopes, and the operators. To quote from the proceedings:

The chairman remarked that about 3 weeks ago, the operations on the mine in question were stopped. A large quantity of water accumulated in the mine, and after a lapse of 10 days was pumped out, and consequently flowed into the river. The river being unusually low, the dead fish had since been traced up to the mine, but above it the fish were in their general condition.

Witnesses were called but no firm proof could be found, so nothing more was done except to call for larger fines in cases like this; the fine for killing one salmon or 1,000 salmon was the same: £5. Although their reasons were mainly selfish, it is interesting to see these early signs of conservation so many years ago.

The mine itself is in private woodland which has no public right of access, and few traces of it exist. The most visible and easily accessible evidence of the existence of this mine is the leat. Starting from Hoo Meavy bridge, it can be traced as a ditch which runs through the wood between the old Tavistock & South Devon Railway (later GWR) track, now a cycle path, and the River Meavy. Around the Goodameavy Bridge area it crossed under the railway track, across the road and back under the track again, at which point there is a sluice for controlling the flow by diverting excess water back to the river below. It then continued through the wood parallel to, and to the east of, the railway tunnel before disappearing into its own tunnel, the entrance of which can be found by following the leat.

Near the far end of the Leebeer tunnel, the leat re-emerges beside the lane which links Shaugh Bridge with Leebeer Farm and is then carried over the railway on an iron aqueduct, very rusty now, which passes high over the track by the southern tunnel mouth. From there it continues parallel to the track but on private land, all the way to Hele Barton. After the mine fell into disuse the leat was used to provide water and power for the farm itself.

Mr Richard Walke, who was brought up at Leebeer Farm, remembered the leat in use. In an interview in 1987, he said:

> At Hoo Meavy bridge there was a sluice controlling the water into the leat, that then goes around the woods and under the railway and eventually goes to Hele Barton. They used it to water the farm and drive a waterwheel to run the barn machinery. They may even have run electric light off it. They had to maintain the leat, making sure it was watertight. They stopped using it in 1953 when the mains water was run in. It used to take two men three days to go along it and clear the leat and repair it. Sometimes animals would break the sides down and it would flood so he was glad to see the leat disused.

RAILWAYS

In the previous chapter, I mentioned in passing the building of the Tavistock & South Devon Railway. Railways played an important part in the history of many of the industries in the valley; without them it would have been very difficult to transport goods around the area. All the industries mentioned in the following chapters relied, to a greater or lesser extent, on the railways, so I feel this is a good place to look at the railways in the valley and the part they played in the development of these industries.

Today we take travel for granted – it is so easy and cheap that we think nothing of it. Many of us have flown across the Atlantic for a holiday in Florida, or to an even more exotic destination. Getting to London is so easy that we can drive or take the train for a shopping trip, returning in the evening. From Plymouth, London is a mere three or four hours away, so driving there and back in a day is no major problem. Getting around Plymouth, even allowing for the awful traffic jams, is relatively quick and easy.

Two hundred years ago travel was a totally different proposition, and the local area would have been unrecognisable. Even Plymouth was nothing like the city we know now; for a start there were several very large tidal inlets which cut into the land. Starting at the northern end of the city, the River Tavy and Tamerton Creek are, of course, still there, as is Ernesettle Creek, which runs all the way to the bottom of Budshead Road. Weston Mill took another bite out of the area, but was not as much of a problem as Stonehouse Creek, which was tidal up as far as Pennycomequick until the end of the nineteenth century, and I can recall seeing the high tide lapping the edges of Millbridge in the 1960s. Millbay was much larger than the modern docks, with what was known as the Sourpool running right up to the modern Millbay Road. The Barbican and Sutton Harbour are more or less unchanged, but there were two inlets from the Laira, the first being a tidal creek plus an area of marshland below Mount Gould, what is now called Tothill Park, and the second being a much bigger tidal creek which ran right up through Lipson Vale to the bottom of Lipson Hill. When added to the huge area of marshland at the mouth of the

Railways

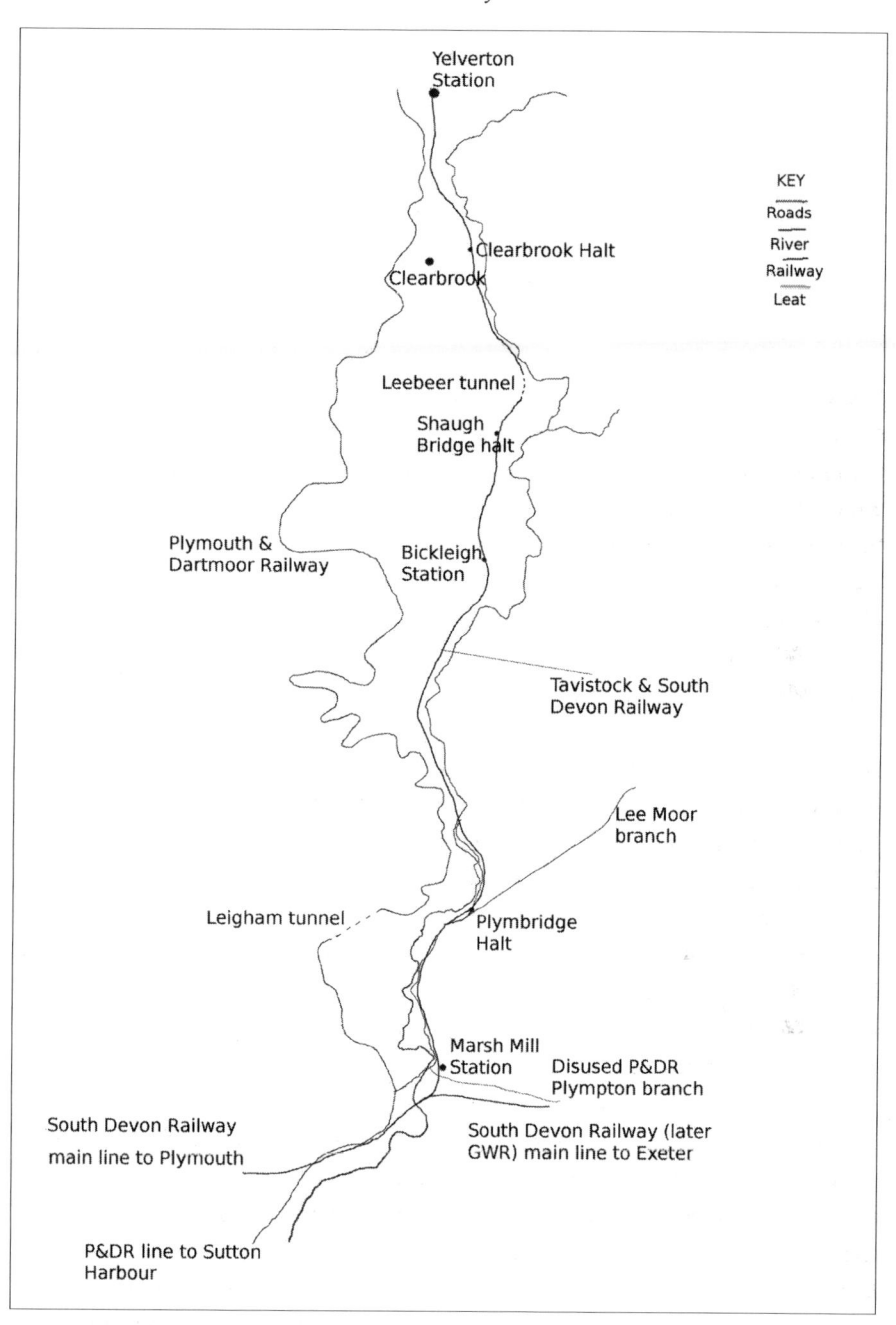

A map showing the railways in the Plym Valley.

Plym which extended up both the Plym Valley and the Forder Valley, it's easy to see how much more difficult travel was in the area.

Plymouth was a major seaport, with links to ports all around the British Isles and the wider world, so getting somewhere by sea was relatively simple. In contrast, however, inland travel was much more of a problem. Apart from the tidal inlets and marshes detailed above, the surrounding area was moorland, riven with numerous valleys which had rivers, streams or marshes in them. The best way to travel was by sticking to the high ground, but this could mean a long, circuitous journey.

A turnpike road linking the city to London had been built in 1758. Having negotiated its way from Plymouth through the marshes and on to Plympton, it then made its way onto higher ground along the Ridge Way, past the coaching inn called the George, and on towards Ivybridge. Even this turnpike would have been little more than a dirt track by modern standards, simply having roadstone and gravel hammered into the local Devon clay, which would need to be re-done regularly as the stones sank into the soil. Given that the average speed of a coach at the time was 4–10 mph, a trip to the capital took at least two, probably three, days and the cost was far out of the reach of working people.

Things had improved slightly by 1841; Edmund, the 2nd Earl of Morley, wrote to his mother that 'you can now very easily get to Exeter the first day' from London, and he regularly spent the night there in the New London Inn (which he recommended), travelling on to Saltram the next day. So by then London was a mere two days' travel away!

In the late eighteenth and early nineteenth centuries local travel was, if anything, even more of a problem. The 1809 Ordnance Survey map shows that many of the 'roads' and walking tracks we know today were in existence, but they were all simply tracks, made by decades or even centuries of usage across the open moorlands and through the wooded valleys. They were made of good old Devon clays, very fine soils which bake hard and crack in dry weather and become slippery, clinging mud in the slightest rain shower. Anyone who walks around the Plymouth area knows only too well that the paths can be a quagmire at the end of a wet Devon winter; imagine all the local roads being like that and you would not be too far wrong.

A further problem is the topography of the area. There are few roads around Plymouth today which offer more than a few minutes of travel on a straight, level surface. Hills and valleys which to us now are a minor inconvenience requiring, perhaps, a change to a lower gear to drive up would have been a major slog on foot or on horseback and even worse in a heavily laden horse and cart.

There would be no improvement in these surfaces until the arrival in this area of the construction methods pioneered by Thomas Telford and John McAdam in the early nineteenth century. Even when these improvements came, they were at first only used on the main roads, with smaller, local roads remaining as bad as they had always been.

Apart from the condition of the roads, the expense of travel meant that it was almost exclusively for the rich. A trip on the mail coach from Plymouth to Tavistock could cost more than a week's wages for a working man. The very rich, the major employers and landowners, had their own coach and a coachman to drive it for them. Minor employers and tradesmen would have kept horses with a cart or trap in which to ferry wives and children. Workers lived as close to their jobs as possible and would walk there and back, often many miles each way, every day. A shopping trip was a rare luxury which meant either a long walk each way or, if you were lucky, getting a lift from a tradesman or carrier in the back of his cart.

To the ordinary working man of the early to mid-nineteenth century, therefore, travel was difficult. Imagine, then, the hardships for the industries of the Plym valley – getting their goods to market would have been extremely difficult. Driving a horse and cart, heavily laden with ore, granite or china clay, along one of these tracks would have been difficult in good weather because of the hilly nature of the area. After heavy rains, it would have been almost impossible.

Other options were, however, becoming available. Since the late eighteenth century, horse-drawn tramways, using flanged wheels running on iron rails, had been used in industrial operations such as mines and quarries. As the tracks were much more level than the roads, and the iron wheels running on rails generated much less friction than cart wheels on mud roads; two horses might pull four or five trucks on a tramway while they could only pull one on a road.

In 1801 Richard Trevithick built his *Puffing Devil*, a steam-powered road locomotive, and in 1804 he adapted one of his creations to run for a short time on rails at Merthyr Tydfil, hauling iron. This was the first steam-powered rail locomotive, although it was much too heavy and inefficient to be used regularly. In 1825, George Stephenson ran his first steam-powered railway between Stockton and Darlington, hauling coal and passengers along a track using wrought iron rails with a gauge (the distance between the rails) of 4 feet 8½ inches, afterwards known as Standard Gauge. Apparently, his choice of this gauge was because 4 feet 8 inches was what was used in the mines; the extra half-inch was given to allow more room to stop the wheels binding on bends.

Various reasons have been given for the choice of this width, starting from the fact that the cart ruts on Roman roads are that size and we have simply carried

Industrial Archaeology of the Plym Valley

A sett used in the Plymouth & Dartmoor Railway track.

A fishbelly rail from the Plymouth & Dartmoor Railway track.

A chair showing a fixing pin.

Railways

A sett showing the imprint of the chair and the holes for the fixing pins.

A pressed steel rail from the Plymouth & Dartmoor Railway alongside a broad gauge rail at the Plym Valley Railway museum. The Plymouth & Dartmoor rail is the smaller one on the right.

A milestone from the Plymouth & Dartmoor Railway.

on building carts and wagons with the same Roman dimensions! As at that time the only motive power was provided by horses, the most likely explanation is that if you start with the back end of a horse, fit shafts around him, then build a cart to fit those shafts and put wheels on the outside of the cart, you end up with about the same spacing between the wheels whether the horse is Roman, Welsh or English.

Railways were thus becoming the preferred option for industrial transport. Even then, the hills and valleys which surround Plymouth made railway building a problem. Railways, whether horse or steam powered, cannot easily cope with steep inclines, so either the railway had to be built along a contour line, meaning that it weaves in and out of every fold in the Earth's surface, or else they went for as straight a line as possible. Given the local topography, this would have meant massive engineering works, constructing cuttings and embankments or viaducts, which obviously meant massive costs. There needed to be a very good reason to build a line, with the hope of a high return on investment, before anyone would take on such a project.

I am deliberately sticking to my brief here, so I will only be discussing railways as they affected the Plym Valley, with only passing mention of where they went outside of the valley.

THE PLYMOUTH & DARTMOOR RAILWAY

The first person in this area to build a railway was Sir Thomas Tyrwhitt. He owned granite quarries at Foggintor near Princetown which provided the stone to build Dartmoor Prison (built between 1806 and 1809) and Nelson's Column (built between 1840 and 1843). He realised that the best way he could get his stone from these remote quarries to be shipped further afield would be by railway, so in 1819 he submitted plans to build a railroad from Foggintor to Crabtree on the Laira. An Act was passed through Parliament the same year for the construction of the Plymouth & Dartmoor Railway (P&DR). The P&DR was opened on 26 September 1823 and initially terminated at Crabtree Wharf on the Laira Estuary, near where the bird hide at Sainsburys is now. The limitations of the tidal wharf at Crabtree soon became apparent, so by December 1825 the line had been extended to Sutton Harbour; it was later extended to a permanently accessible wharf at Cattedown near Laira Bridge as well.

The P&DR was a track of the first kind described above, in that it followed the contours. Although Sir Thomas called it a railway, this could be misleading; we tend to think today of railways as using locomotives to pull passenger-carrying trains when it was in fact a horse-drawn tramway carrying only goods. From the point of view of the Plym Valley, it ran from Crabtree across the marshy Forder Valley and through a tunnel in the Leigham Ridge (only the second such tunnel to have been built in England at the time), bringing it into the Plym Valley. From Leigham it wound its way high above the river, through Mainstone, Bridgett and Rumple Woods to Riverford. After Riverford, it left the Plym Valley and continued through Common and Fancy woods towards Roborough, at that time known as Jump. This meandering route took nearly 9 miles to cover a straight-line distance of 4.

From Roborough it meandered on across Roborough Down towards Yelverton and then on towards Princetown. The railway was originally made from cast-iron rails, known as 'fish-belly rails' because of their curved underside, fixed to cast-iron chairs which were supported on individual setts or stone sleepers. These were generally made from granite, although some near Rumple are slate because that was the cheap, plentiful local stone. The setts varied in size; many were roughly 2 feet square and 1 foot deep although some, including the unearthed one pictured, were 2 feet wide by 1 foot square. Given the overall length of the railway at 25 miles, that means roughly 88,000 setts were needed and the production of these alone must have taken a huge amount of the output of the quarries. Most remaining ones are set at 1-yard intervals, although occasionally there are sections with different spacing. The fish-belly rail which is in the Plym Valley Railway museum is 3 feet 2 inches long overall, with 2-inch tongues at either end, although one is partly missing. These tongues were tapered so that two from consecutive rails could both fit together inside a chair and give the distance between chairs of 3 feet.

Near the eastern end of the Leigham Tunnel is an area where many setts are visible, and here there are intermediate setts between the originals, making them almost continuous. Some here also have a groove worn across them; others have an imprint of the rectangular area where the chairs were fixed, not necessarily aligned with the later groove, so they were obviously recycled from elsewhere. These, and the unevenly-spaced setts mentioned above, must have had the later type of pressed steel rail fixed directly to them. An example of both the fish-belly rail and a length of the pressed steel type can be seen at the Plym Valley Railway museum.

The track had a siding every mile or so, where it was appropriate, to allow for passing as well as for loading materials. There was a loading siding at Higher Rumple Quarry, and on the hillside below this there are two large blocks of granite, one 4 feet square by at least 2 feet deep, the other 6 feet by 4 and at least a foot deep. Were they platforms for loading derricks, dumped after the railway was broken up, or the evidence of some accident where they fell from a wagon and slid down the slope? We shall never know. Certainly there is no granite locally, so they were definitely brought here.

The railway had round granite milestones with a number cut into the top showing how many miles the particular point was from Sutton Pool, the terminus. A few of these can still be found in place today.

Slate setts at Cann.

Railways

The spacing between the rails, known as the gauge, was 4 feet 6 inches, which became known as Dartmoor Gauge, and all the branches of the P&DR used that gauge. No viaducts were constructed; in the Plym Valley, the railway was mainly built by cutting a flat ledge into the side of the hill and, where necessary, constructing an embankment.

Tyrwhitt had always hoped to carry goods for other businesses, apart from his quarries at Foggintor, in order to help defray the costs. He had intended the returning wagons to take coal as well as lime and fertiliser to improve the moorland soil, and it had also been his intention to carry tin, copper, china clay and slate for others – at a price of course!

The railway's route took it past the existing Rumple and Higher Rumple slate quarries in the Plym Valley and would have vastly improved their transport links. At Higher Rumple the railway ran across the entrance of the quarry, but for Rumple quarry it meant transporting the slate up the zig-zag path from the quarry to the railway line above, which must have been very difficult. The enormous effort required was, however, much easier than any alternative and thus worthwhile. Colwill and Hyren quarries opened some time after 1839 (they do not appear on the Tithe map of that date), and both made use of the transport link.

Colwill fronted directly onto the line, like the adjacent Higher Rumple, so their slate could be loaded directly on to wagons at the mouth of the quarry. At Hyren quarry, near Riverford, the operator built a steam-powered inclined plane to carry the stone some 200 yards directly up the hillside. When Mr Henry Crace, the quarry's operator, sold up in September 1847, the auctioneer's advert included the following:

> Lot one will consist of all that superior steam engine 15 horse power with boiler fixed complete with copper piping ditto now erected on the above quarry together with carts and other apparatus thereto belonging. The steam engine is quite new and has been recently erected on the quarry and is of the best material. In addition to its being fitted for raising the slate from the quarry, it has all the requisites for being used as a saw mill, having a circular saw bench complete with machinery.

The inclined plane itself is shown on the 1853 deposited plan for the later Tavistock & South Devon Railway, and evidence of the upper section is still visible between that railway and the P&DR.

Cann quarry was across the river and thus had no easy access to the P&DR. However, when the P&DR was being built the directors of the company wanted to pass over some of Lord Morley's land. Part of the deal agreed privately by

Industrial Archaeology of the Plym Valley

The Lee Moor Tramway, showing a passing section. (*CCHS*)

Lee Moor engine No. 2. (*CCHS*)

The bridge for the Lee Moor Tramway over the road at Plymbridge in the early twentieth century. (*CCHS*)

A view down the Lee Moor Tramway showing the accident site. (*CCHS*)

three of the directors on behalf of the company was that the P&DR would, at their expense, connect Cann quarry (which was owned by Lord Morley) to their tramway, also 'procuring a responsible tenant for the said quarry for a term of fifty years at a clear fixed annual rent of 150 pounds over and above any further tonnage' to rent the quarry! As the tramway at its nearest is 100 feet above the quarry and on the other side of the valley, this was a ridiculous deal to have agreed. It would have been a very tricky and expensive operation, if it was possible at all. When faced with the problems, the company initially refused to comply with the agreement; a report dated 6 July 1824 states that it is 'quite out of their power to carry out the same into effect'. After two of the three directors who had agreed to the deal were bankrupted by the collapse of the Plymouth Bank in which they were partners, leaving John Pridham alone to face a possible £5,000 personal claim for which the company was in no position to indemnify him, the company caved in and agreed to fund whatever was needed.

Initially, Lord Morley preferred using a canal to carry his slate to Crabtree, and the leat which served the Marsh Mill was duly widened and strengthened between 1827 and 1829, when the canal started carrying stone. The canal was short-lived; possibly by 1835 and certainly by 1839, depending on which source you believe, the P&DR was extended from the Marsh Mill junction to Cann

A horse and wagon of the Lee Moor Tramway beside the main line at Laira. (*CCHS*)

along the towpath of the canal, which reverted to being simply the mill leat again. Here the setts, not surprisingly, were made from slate and are much more irregular in shape than the granite ones.

A short-lived branch from the P&DR line had been constructed from Crabtree to Marsh Mills and on to Plympton in 1833 to carry china clay to the docks. This clay had been carried down from Lee Moor by pack horses and stored in a warehouse near St Mary's Bridge, from where it was loaded into trucks on the P&DR line. This branch closed in 1847 to make way for the South Devon main line, although its original track is still in existence as the raised footpath which runs alongside Plymouth Road almost as far as St Mary's Bridge.

In 1858 Lord Morley again seems to have done very well; this time, in exchange for allowing the use of some more of his land, he persuaded the Tavistock & South Devon Railway to pay for a branch which ran from the Cann Quarry branch directly to the Lee Moor china clay works, which was also on his land, 'providing that the Dartmoor Gauge was used on the lines; if for any reason it was replaced then this agreement would become void'. This track snaked across the moorland from the clay pits and crossed the road from Shaugh Prior to Plympton near the present-day Cann Wood car park via a gated crossing. The tracks were removed in the late 1960s; I can remember

Railways

Setts at Clearbrook.

The Whitegates crossing near Woodside with the rails still in place. (*CCHS*)

Bridge supports beside the Forty Foot Drive.

Industrial Archaeology of the Plym Valley

The control wheel for the inclined plane beside the Forty Foot Drive.

The Lee Moor Tramway bridge at Plymbridge today.

Railways

Weighbridge Cottage and the iron bridge.

seeing the white gates of this crossing still in existence in the 1980s and the support posts for them are there still.

From there it went down through Cann Wood via a long inclined plane, running much of the way through a cutting which in places is 20 feet deep, crossing Plymbridge Road on a bridge and joining the Cann Quarry branch about 100 yards below Plymbridge. The upper section was later steam powered, having two small saddle-tank 0-4-0 locomotives named, imaginatively, *Lee Moor No. 1* and *Lee Moor No. 2*, which delivered the clay from the pits at Lee Moor to the upper end of the inclined plane. Gravity provided the power on the inclined plane, the loaded wagons coming down the line being much heavier than the returning ones, even if the latter were loaded with supplies and equipment. The speed of the wagons was controlled by a brake on the winding-drum at the top of the inclined plane. For most of its length the tramway had only two rails (unlike the Dewerstone incline – see Dewerstone Quarries chapter) but halfway down there was a passing section where the two rails became four separate rails so that the up-going and down-going wagons could safely pass each other, and beside this passing section there was a corrugated iron hut (the remains of which are still there!), presumably for a watchman to work the points and make sure all was well. The line crossed over Plymbridge Road on a trestle bridge. From Plymbridge onwards, the line was horse-drawn; in 1904 there were discussions with the GWR about making it steam powered but nothing came of them.

There were times, of course, when things went wrong. In 1865 the newspapers carried the following story:

LEE MOOR TRAMWAY ACCIDENT
September 4th 1865
(LOCAL INTELLIGENCE)
A serious accident occurred on Friday last on the tram road of the Lee Moor Clay Works. One horse was killed on the spot, a man seriously injured on the head, four trucks were smashed, some being thrown on end and some having the wheels and axles broken, while the clay bricks with which they were laden were scattered in all directions. The noise was heard 2 miles away at Plympton Bridge [presumably they mean Longbridge on the Plympton road]. The tram road is about 1½ miles long, and the incline portion is very steep. The laden trucks are let down with a long rope which passed round a drum at the top of the hill, the other end being connected with the empty ones [trucks] at the bottom. The laden trucks descend by their own weight and pull the empty ones up, the velocity being checked by the brake at the top. The full ones had descended about half way when the coupling iron of the first truck broke, and the rest descended at a frightful speed, so fast indeed that the eye could hardly distinguish them. The man was standing on the level part below with 2 horses, waiting to take the trucks on after their arrival. He at once perceived what had happened, and took the two horses into a shed which is on the siding, expecting the truck would keep on the main line, which had a curve at that spot. The curve however was doubled up, and the wagons rushed into the shed, one over the other. The man saved one of the horses and was in the act of saving the other when he was knocked down in the attempt.

The line constructed to carry granite was now carrying slate and china clay as well, but despite all the goods which were carried on the line, it is unlikely to have ever made a profit. The initial cost of building the line overran by nearly 50 per cent; it cost £66,000, although it had been estimated at £45,000. The legal disputes further depleted the company's reserves and as many of the major shareholders were also major customers, they seem to have used it for purely selfish means, insisting on the very lowest fees for their own businesses, which ensured that the income never covered the costs. The company limped on until the Plymouth–Yelverton section closed late in the 1870s; some of the Yelverton–Princetown trackbed was rebuilt to take standard gauge steam engines in 1893. This line closed completely in 1956. The Cann Quarry branch was used occasionally up until the final operator retired in 1910. The Lee Moor branch closed in 1939, the rails being replaced by a pipeline which delivered the clay slurry to a drying kiln at Marsh Mills, but the stretch from Marsh Mills to Laira Wharf was in use until the mid-1960s, carrying china clay blocks from the drying kiln to the wharf, and many pictures survive showing the incongruous-looking clay-white horse-drawn wagons crossing the main lines.

So, what can be seen of the Plymouth & Dartmoor Railway today? The Plymouth end, from Cattedown and Sutton wharves to Crabtree and across Forder Valley, has mostly been buried under new roads, although occasionally stretches of the trackbed can still be made out if you know where to look. It used to pass around the Efford Promontory, near where the Beefeater pub is now. It was there that the spur which originally served both Cann Quarry and Lee Moor joined it, having crossed the river parallel to the old road on the Long Bridge. From this junction, it originally ran diagonally across the marshy ground to the east side of Forder Valley. There is a section of trackbed in the Forder Valley Nature Reserve, now used as a footpath running from the entrance to the Parkway Industrial Estate, off Leigham Roundabout, along past the original 3 mile marker stone (signifying 3 miles to Sutton Wharf) to the southern portal of the Leigham Tunnel. Leigham Tunnel is blocked and inaccessible, and some sections of the trackbed between there and Yelverton are now on private land.

Other sections along the route are accessible but are frequently overgrown and/or muddy. The most accessible part is the section above Clearbrook, between there and Yelverton, much of which has been turned into a cycleway. There are long sections here where the granite setts are very much in evidence, and of course there is Sir Thomas Tyrwhitt's original stable beside the track near the ninth tee of Yelverton Golf Course, now used by the club as a store.

The path which runs from the National Trust car park at Plymbridge between the canal and the River Plym to Cann Quarry is the remains of the Cann Quarry tramway, and many of the slate setts are still visible, although no trace of the track or fixings remains.

Sections of the Lee Moor branch can still be seen; it crossed the Plympton to Shaugh Prior road at a place called Whitegates, between the entrance to Brixton Barton and an unnamed house on the right as you drive towards Plympton. The name came from the colour of the crossing gates, which were still in existence until the late twentieth century but were removed several years ago. The large steel pillars on which they were fixed remain beside the road, near the unnamed house.

From this crossing, the track meanders northwards towards the Lee Moor works, well outside my area of interest, but heading southwards, it carries on through Cann Woods and the inclined plane section and its cutting are easily found. At the top of the incline the winding house and its wheel are still visible, constructed within an embankment beside the Forty Foot Drive, one of the main tracks through the woods. The tramway used to cross over this track by a bridge, the supports for which are still there.

At Plymbridge can be seen the most prominent remaining structure: the bridge across the road which carried the tracks until 1939, and the pipeline thereafter. Sadly, gone is the old wooden structure shown in the previous picture of this bridge; the supports are now simple stone-built pillars with a small steel platform on top which carried the pipeline. A hundred yards further on, near the southern end of the National Trust car park, is the old stable block where the horses were kept, the site of the accident of 1865 described above. A hut for the man in charge of the horses is built on the road end of the stables, with a fireplace and even what looks like an oven. Also, here the two tramways crossed over the Cann Quarry canal/leat by a pair of bridges, now very rotten and dangerous.

There are many granite setts along the length of the inclined plane, but existing photos of the track, albeit from the early twentieth century, show wooden sleepers so presumably the early fish-belly rails on these setts were replaced with pressed-steel ones on these sleepers at some stage. At the junction where the tramway joins the Cann Quarry branch there are some sections of this pressed-steel rail if you know where to look.

There are still occasional sections of the steel hawser which hauled the wagons up the incline to be found, about 1 inch in diameter, and also sections of the later 10-inch-diameter steel pipeline and a much later 4-inch-diameter plastic pipe, still with traces of china clay inside.

Cann Wood, owned by the Forestry Commission, has an open-access policy, but the track of the tramway is still owned by Imerys, so technically nobody should walk on it. It is very overgrown with many fallen trees blocking it, especially in the cutting on the inclined section, and anyway everything can be seen from the paths which run beside it so there is no real need to walk there.

At Marsh Mills the line crossed the marsh beside the Long Bridge, but it first had to cross the River Plym. It did this on a cast-iron bridge, which is still in existence beside Weighbridge Cottage. Here it is also possible to find stretches of original rail still in place on the footpath which runs from the Coypool Park and Ride to Longbridge Road.

The Tavistock & South Devon Railway

In 1853 the Tavistock & South Devon Railway Company put forward a plan to build a railway along the Plym Valley to Yelverton and thence to Tavistock. Lord Morley was involved in this railway as well, having sold the company some of his land for the construction of the track and buying shares in the company in exchange for their agreement to build the Lee Moor branch line, as described above. The Act of Parliament was passed in 1854 and construction started on this mammoth engineering feat, requiring the construction of four massive trestle-type wooden viaducts between Plymouth and Yelverton alone, at Cann, Riverford, Bickleigh and, the longest at 190 yards, at Ham near Shaugh Bridge. Also needed were smaller stone arch bridges at Plymbridge, Goodameavy and Clearbrook, and a tunnel some 1,000 feet long at Leebeer, ending just before Goodameavy bridge.

The original chief engineer on this project was Augustus H. Bampton, but when he died in 1858 Isambard Kingdom Brunel took over. The T&SD was the second type of railway referred to previously, in that it ran as straight as it could, allowing for the terrain. Work started in 1854 and it took five years to complete, opening on 22 June 1859. From its junction at Marsh Mills with the South Devon Railway, which linked Exeter and Plymouth, the T&SD line was constructed along the line of the valley, crossing the river between Cann and Rumple quarries. Although this did not directly affect either quarry, the work necessary to build the line and especially the viaduct required some very complex engineering and changed the area almost beyond recognition. On the Cann side of the river, the railway was built along a ledge cut into the hillside and reached the crossing point at an area where spoil had been dumped for many years, so the ground was not stable. A small pier was built to take the railway as far as possible past the pre-existing cottages. The first support on the Cann side was a simple wooden trestle with a single diagonal support.

On the Rumple side, the railway would disappear into a cutting; an embankment was built from the cutting, also in an area where spoil had been tipped, jutting out toward the river. Five stone-built supports were constructed, three simply being ground-level reinforcements made as stable as possible sitting on bedrock, and the central two on either side of the river being tall plinths built on rock.

The timber framing for the viaduct was a wondrous piece of construction; it was built piecemeal by lowering beams, some weighing half a ton, which had been previously cut and drilled to Brunel's design, from the embankment above and securing the lower end into a shaped iron casting fixed to the top of the support. The structure was then temporarily braced and the load-bearing road built on top, allowing the builders to move out on that and lower the next beams into place.

The fan-shape of the trusses and the fact that they were used throughout Cornwall gave rise to the name 'Cornish Fan viaducts'.

All this construction, which nowadays would be a simple operation involving a few men and some powerful machinery, was completed using nothing but manpower and winches, an amazing feat. This process was then repeated three more times between there and Yelverton, and twice afterwards, at Bedford Bridge (colloquially known as Magpie) and Walkham.

Several areas required long cuttings to be made, sometimes 20 feet deep, through the stone of the hillsides; others, while not requiring viaducts, needed substantial embankments to be built. All in all this was a huge effort, which changed the landscape of the Plym Valley forever; in fact, it is difficult sometimes to envisage how the valley would have looked before all this work.

The T&SD joined the mainline South Devon Railway at Marsh Mills, and like that railway the Tavistock & South Devon used the Broad Gauge track, 7 feet and ¼ inch wide; the Plym Valley Railway museum has a section of broad gauge rail which can be seen.

So now we had two railway systems with mutually incompatible gauges running through the valley, which meant that everything which was brought to the T&SD from the P&DR would have had to be offloaded and reloaded each

Brunel's original Cann Viaduct with the Cann quarry behind. (*SSPL*)

Railways

time. To complicate things further, from October 1874 the London & South Western Railway (L&SWR) built their own line from London to Plymouth which ran north of Dartmoor, came down from Okehampton and joined the T&SD around Tavistock. They gained permission to run down through the valley along the existing trackbed. The L&SWR used the Standard Gauge of 4 feet 8½ inches, so from Tavistock down to Marsh Mills a third rail was added to the existing line, between the other two, in order that both Broad and Standard Gauge engines and rolling stock could use the line. This system was known as Dual Gauge.

The P&DR line between Yelverton and Plymouth was largely disused by the 1870s, but the Cann Quarry/Lee Moor branch was still working. This meant that for a short time the stretch between Plymbridge and Marsh Mills had three mutually incompatible gauges running in parallel!

On New Year's Day 1876, the Great Western Railway (GWR) took over the running of the T&SD. The L&SW stopped using their route in 1890 and in 1892 the GWR converted the broad-gauge lines to standard gauge.

The design of Brunel's wooden viaducts was such that repairs and replacement of timbers were relatively easy, but they were showing signs of age by the end of the nineteenth century so the GWR decided to build new stone ones alongside the old ones. Riverford and Bickleigh were the first to be replaced, the new ones coming into use in 1893, and Cann was the final one in this section, finished in 1907. The instability of the spoil on which Cann viaduct was built has already been mentioned, and because the rigid stone viaduct would be less able to cope with any slight movement than the much more flexible wooden one, the new viaduct here required some extra bracing to make sure it was strong enough. This included filling in the second arch on the Cann side with bricks and leaving in place the wooden former used in constructing the small first arch on that side. This former was still there in the 1980s when I first knew the area, but eventually some moron thought it would be fun to set fire to it, and thus was lost another piece of history. The cracks in the brick infill of the second arch, easily visible to anyone walking down the slope towards the quarry, show just how much the viaduct has moved over the years.

In order to minimise disruption to the service during the building process, the new viaducts were built alongside the existing ones and realignments carried out to cuttings, etc. at their ends, with the track on the old viaduct still in use. Riverford is a good example of the extra complications that might be involved when this work was done. The wooden viaduct was built on a curve, with the railway approaching through cuttings at each end. The track ran centrally through the fairly narrow cutting at the southern end, so as to make room for the realignment the cutting needed to be widened. A hundred yards before the start of the viaduct, a stone arch bridge carried one of the tracks from Riverford hamlet up through Common

Wood to the Plymbridge Road, and the arch of this bridge was too narrow to allow enough room for the realignment. The only option was to build a new, iron bridge beside the stone bridge, then demolish the existing bridge over the cutting (the remains of the original arch bridge can still be seen on the river side of the cutting) and, finally, widen the cutting.

The new stone viaduct was built beside the wooden one, on the side furthest from the river, with only a few feet between them. Once all this work was complete the track was laid across the new viaduct and finally joined to the existing track, allowing the service to continue without a hitch. The wooden viaduct could then be demolished piece by piece, as it had been constructed, leaving just the three stone support pillars in place.

During the course of the track's useful life several stations, platforms and halts were built. Between the mainline junction at Marsh Mills and Yelverton, there were the following, in order of passage along the line:

- Marsh Mills station opened 1861
- Plymbridge Platform, opened 1906
- Bickleigh Station opened 1859
- Shaugh Bridge Platform, used as a siding during the building of Leebeer tunnel, reopened as a siding for iron ore in 1870 then as a platform for passenger traffic in 1907
- Clearbrook Halt opened 1928
- Yelverton Station opened 1885

(All this information thanks to Bernard Hill's book *The Branch*)
These stops would originally have been opened for the benefit of local inhabitants and industries, including the china clay from Shaugh Bridge kiln, which at first was carried all the way to Bickleigh station, but was later taken to Shaugh Bridge siding, and iron ore from Shaugh mine, which also went to Shaugh Bridge siding for onward carriage. Many, however, came into their own during the first half of the twentieth century, when Plymothians wanted to visit the surrounding glorious countryside. Anywhere along the valley could be visited for *6d*, so they were known as the Woolworths' Specials after the store, which at the time sold its wares for that price. People could travel from Plymouth to Plymbridge platform or Shaugh Bridge platform, then spend the day walking in the beautiful countryside nearby, catching the train home in the evening. Several local people made good money supplying tea and snacks for these visitors, and just travelling along that beautiful valley by steam engine must have been a real treat.

With the proliferation of private motoring and improvements to roads in the 1950s and 1960s, rail travel became less popular and, treat or not, if it did not make enough money it could not survive. As with much of Britain's

Railways

A wooden reinforcement under the arch of the Cann Viaduct. (*John Luscombe*)

An arch of the Cann Viaduct showing brick reinforcement.

Industrial Archaeology of the Plym Valley

The iron bridge near Riverford Viaduct. The remains of the stone bridge are on the right-hand side.

One of the original supports for the Cann Viaduct with the new viaduct behind.

Plymbridge platform. (*CCHS*)

Railways

Bickleigh station. (*CCHS*)

Shaugh Bridge platform. (*CCHS*)

railway system, the economics were closely studied by the infamous Dr Beeching. In 1962 the passenger trains were stopped and by 1964 the tracks were demolished.

So what remains from this massive engineering feat? Surprisingly, a great deal, because much of the trackbed has become a foot and cycle path. Even more surprisingly, the Marsh Mills station platform is still in existence and can be found between the two approach roads which lead to the Currys and B&Q industrial estates, with a section of the old railway line running unused beside it. The cycle path can be accessed from the Park and Ride at Coypool and from there to Plymbridge

Industrial Archaeology of the Plym Valley

it follows the trackbed of the old Cann Quarry and Lee Moor tramway, running parallel to the GWR track. The original Plymbridge platform is no more, although the Plym Valley Railway Co. Ltd has relaid the GWR track from Coypool to Plym Bridge and built a new platform to run trips along this part of the line as a tourist attraction.

The cycle path from Plymbridge then follows the old GWR/T&SD track as far as Bickleigh Station, which is now privately owned, but a new section of cycle path has been built, bypassing this; the path then carries on past the remains of Shaugh Bridge Halt, where the platform is still visible, on through Leebeer Tunnel and as far as Clearbrook, where again the halt is now privately owned and the path ends.

The path runs through some glorious countryside and often gives a unique view of some well-known places. All the viaducts are still in place on this section; only the later Walkham viaduct between Tavistock and Yelverton has been demolished, although this has recently been replaced by a multi-million-pound bridge for cyclists. The views from some of these viaducts can be superb, especially that from Cann, which straddles the river and overlooks Cann Quarry, home for many years to peregrine falcons, ravens and kestrels.

The Cann Viaduct today.

DEWERSTONE QUARRIES

The Goodameavy Estate is an area of land between Clearbrook and Shaugh Bridge based on Goodameavy Manor. The estate extends along the east bank of the River Meavy from Goodameavy Bridge as far as Shaugh Bridge and up on to the moor as far as Urgles Cross, and includes the large granite outcrop around the Dewerstone, much loved by climbers, and Carrington Rock. Granite is the bedrock in the area, as is obvious to anyone walking in the area because there are huge quantities of large granite boulders on the hillside, known locally as moorstone. These boulders had been used for centuries as building material, with large boulders being hauled away, cut roughly to shape and incorporated into buildings, especially as door and window lintels.

The estate was bought from its previous owner, Joseph May Ward, by Joseph Scobell in 1815. Mr Scobell set about making improvements to the area; he built a new bridge across the river below his manor house, celebrated by a stone set in the south-eastern side of the bridge wall with his initials and the date 1832 still clearly legible. He is also said to have built a new road from the western side of this bridge to join the road which connects Clearbrook to the main road (now the A386) across Roborough Down. In fact, there was already a track going in that general direction shown on the 1809 OS map, so I assume he must have improved and re-aligned it rather than built it. It now joins the Clearbrook road just below the top of the hill, beside the bridge where the road crosses over Drake's Leat. When Joseph died in 1836, the estate passed to his son Edwin.

Edwin and his wife Georgianna owned the estate when, in the late 1850s, they were approached by the Johnson Brothers, John and William, who at the time were operating granite quarries at Foggintor on Dartmoor and also running the Plymouth & Dartmoor Railway; in both these operations they had succeeded Sir Thomas Tyrwhitt. In the late 1830s they had bought a seven-year lease of the more or less worked-out Haytor quarries near Ilsington on south-east Dartmoor from the Duke of Somerset. Haytor granite had a name as being stone of excellent quality. The Johnson Brothers then closed Haytor down and transferred the name 'The Haytor Granite Company' to their existing quarries near Foggintor. The

Industrial Archaeology of the Plym Valley

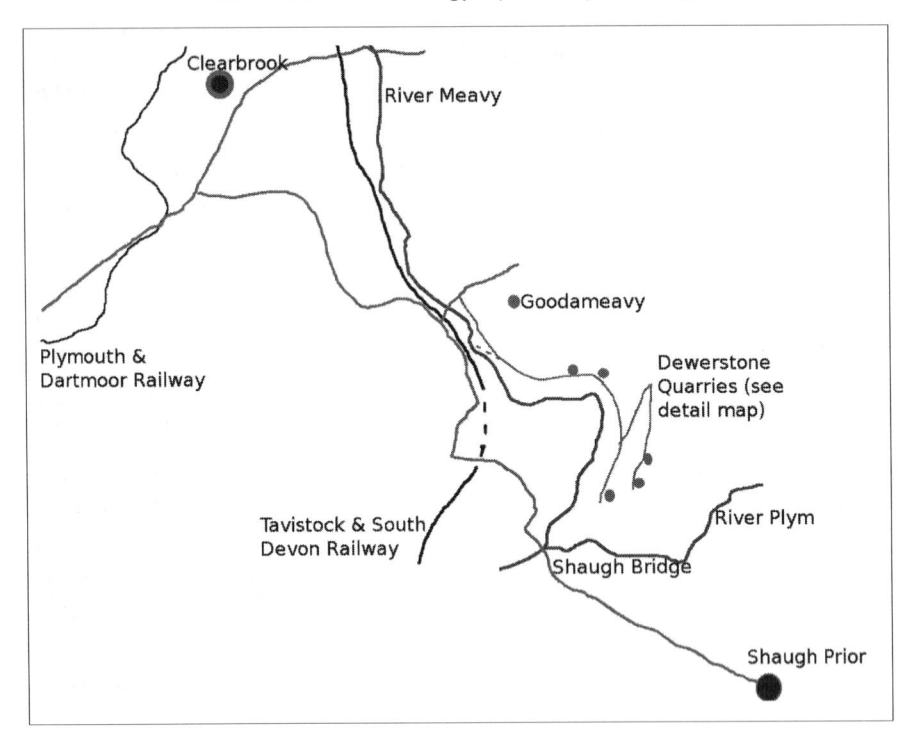

The location of the Dewerstone quarries.

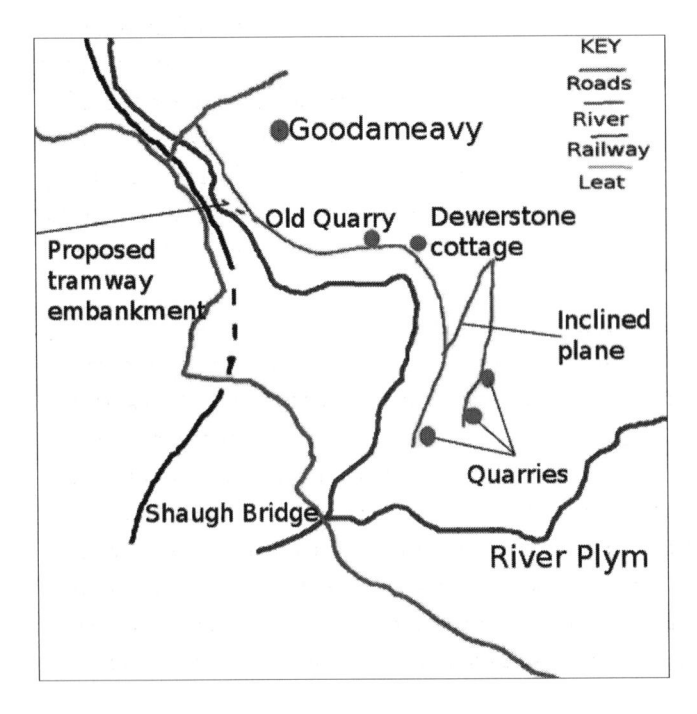

Detail showing the Dewerstone quarries.

54

Moorstone.

more I read of their activities at the time, the more I feel that, if they were not actually doing anything illegal, at the very least they might be considered guilty of sharp practice. Anyway, in 1857 the Johnson Brothers contacted Edwin Scobell to investigate the possibility of working the granite outcrop around the Dewerstone.

Letters exist from around this time between the Johnson Brothers' company and the Tavistock & South Devon Railway which show their intention of building a siding at Goodameavy where they could bring their stone for direct loading on to the railway. One letter dated 2 November 1858 states:

> I am desired by the directors of this company to make application to the directors of the South Devon & Tavistock Railway Company [sic] to be permitted to make a siding near Goodameavy Bridge to connect a branch railway now in course of formation at Dewerstone with the South Devon and Tavistock Railway as shown by the enclosed tracing. The Haytor Granite Company have had leased to them by Mr Scobell the right of quarrying granite at Dewerstone and of making the branch referred to. So far as Mr Scobell and Sir M. Lopes are concerned and at several interviews which I have had with Mr Carr, the secretary of South Devon Railway Company respecting the transit of the granite, it has been stated to me that at the

proper time, the directors of his company will be ready to make such arrangements as without doubt will be mutually satisfactory. It is the intention of the directors of the Haytor Granite Company *to open quarries at Dewerstone* [my emphasis], having proved to their satisfaction the quality of the rock there and work them in conjunction with the quarries on Dartmoor.

From this letter, it can be seen that the quarries opened around this time and also that neither Edwin Scobell nor Sir Massey Lopes, the two landowners, had any objections to the construction of the proposed siding. Another letter below seems to show that all parties, including the railway company, were happy for the siding to be built.

December 13th 1858

To the South Devon and Tavistock Railway Company Mr H Evans Esquire, Secretary

Sir,

I am desired by the directors of the Haytor Granite Company to acknowledge receipt of your favour of the 19th ultimo, enclosing the extract of the minutes of the proceedings of your board in regard to my application of the 2nd ultimo to request that the agreement be prepared as proposed and for the same term as the lease of the Dewerstone which expires at Michaelmas 1884 with proviso that if the Haytor Granite Company shall renew their lease of Dewerstone that the South Devon and Tavistock Railway Company shall renew their agreement with them on the same terms as before.

I am Sir

Your Obedient Servant

C Hoar, Manager

A cutting was made through solid rock, and then an embankment some 200 yards long was built to connect the tramway from the quarries to the proposed siding across the river; the embankment even included an arch bridge in its length to allow an existing riverside track to pass through. Also constructed was an area of flat ground adjacent to the railway itself, presumably to allow

Dewerstone Quarries

The bridge through the embankment to the Meavy River.

room for the siding, and a stone support for the bridge, both of these works being on Sir Massey Lopes's side of the river. For reasons which are unclear, the connecting bridge across the River Meavy was never built. It seems ridiculous to have done such an enormous amount of work but not put in the final piece.

The popular story about this debacle says that Sir Massey Lopes changed his mind after they had done all that work and refused them permission to cross his land. Given the documentary evidence quoted above, plus the fact that he had already allowed them to build the support for the bridge on his land, this is obviously untrue. They had the opportunity simply to trundle the granite down the tramway, across the river and transfer it to the waiting wagons of the railway, which could then deliver it to anywhere in the country. Many businesses in such remote locations would love to have such easily accessible transport links, and having already built a long embankment, it seems insane not to have built the final linking bridge.

The real reason for not finishing the tramway, I feel, lies in the fact that the Johnson Brothers had good reasons for preferring their existing transport arrangement. This would have been to cart the stone along the existing track to Goodameavy Bridge and then up the hill to join the Plymouth & Dartmoor Railway above Clearbrook. As the Johnson Brothers owned that railway, their costs from that point would certainly have been cheaper; in 1834, in exchange for relinquishing their monopoly on the use of the railway, they had negotiated a reduction on the already ludicrously cheap payment per ton for their granite from 2s 6d to 1s 10d. The effort required to get the granite from Dewerstone quarries to Clearbrook was huge given the state of the roads at that time, which would have been little more than dirt tracks, but presumably they could not agree with the South Devon & Tavistock Railway a price which matched their costs on the Plymouth & Dartmoor and they put financial saving above all else; the bridge was never built.

Whatever the reason, the quarries appear to have been working and producing granite at this time and the unused link to the siding at least improved access to the road at Goodameavy, allowing the stone to be transported that way. Tramways were built joining the quarries, employing what we now call Standard Gauge (4 feet 8½ inches). Over the years people have said that the rails of these tramways were either narrow gauge or Dartmoor Gauge, like the P&DR, or even a mix of gauges, but there are enough setts with the original fixing holes still visible on each level to check the measurements between them. The distance between the centres of the rail fixings is 59 inches, almost exactly the same as that of modern rails, which would give a gauge, i.e. the distance between the rails, of 4 feet 8½ inches; they were definitely Standard Gauge.

The tracks were fixed to granite setts rather than wooden sleepers, just like the P&DR tracks, but they appear to have used the later pressed-steel rails rather than the 3-foot-long cast-iron rails originally used on the P&DR.

The quarries were on two levels, with two larger quarries close together on the upper level and two smaller ones more widely spaced below. One of these lower quarries is marked on the 1885 OS map as 'old quarry' so it was at that stage disused, possibly because either it had been simply a trial dig or else it was not very productive. Each working quarry had a form of crane on a circular platform to help with loading granite onto wagons on these tramways. According to Dr Peter Stanier, who has done his PhD on granite quarries, these cranes were probably of the type known as 'spider cranes', which had a single timber mast with six or more stays from the top of the mast anchored to the quarry walls like a spider's web. A jib fixed to this mast could then turn easily in any direction, unhindered by back-stays, and an associated hand-winch could be used to lift and move the stone.

An inclined-plane tramway was also built down the slope, joining the upper and lower tramways. This had only three rails for most of its length, the wagons on both sides using the central rail. For about 55 yards in the middle of the inclined plane, the central rail was split into two to allow the trucks to pass each other. The two wagons were joined by a long cable which went up

Granite setts on the tramway.

from one, around a wooden drum in the winding house at the top, then back down to the other truck. The winding house controlled the operation, using the weight of a wagon full of granite both to lower itself down the inclined plane and to haul up the empty wagon on the other track, plus any equipment needed. No motive power was needed; the drum in the winding house simply had a steel belt around it which could be tightened to act as a brake.

The lower tramway must have extended past the inclined plane at least to Dewerstone Cottage, which was then the counting house, smithy and stable, if not further along to the embankment. A hundred yards past Dewerstone Cottage is the disused quarry previously mentioned. Horses would have been used to draw the wagons along the level tramways.

The quarries were operated by Johnson Brothers for around 20 years. By 1878 Mr Scobell was obviously sick of the Haytor Granite Company, and glad to see the back of them. In a letter to Sir Massey Lopes, he says 'As I find the wayleave for granite you so kindly granted to me many years ago expires in 1884 I shall be much obliged if you will now give me a promise of an extension. I am free of the Haytor Company and about to advertise the quarries and although your permission to cross the river has never as yet been taken advantage of, I hope a fresh company may be glad to avail themselves of it.' Again there is mention of

Industrial Archaeology of the Plym Valley

A circular platform for a loading crane at the entrance to the quarry.

Dewerstone Quarry winding house in about 1910. (*NT*)

the wayleave and permission to cross the river being agreed between the two landowners. At about this time the Plymouth–Yelverton section of the P&DR closed down, so that form of transport would no longer be an option. Despite this, the 1885 and 1907 OS maps show the embankment and even the levelled area beside the railway, but no bridge across the river. The obvious, commercially sensible link was never built.

There is no record of who, if anyone, took on the operation of the quarries at that time, but Mr Scobell carried on paying wayleave of £20 per year to Sir Massey on behalf of the quarries, so presumably someone was working there either on Mr Scobell's behalf or working for themselves and paying Mr Scobell's lease. Following the death of Mr Scobell in 1884, the wayleave continued to be paid by his widow until March 1885. In those days, £20 was a large sum of money and while it might be that the agreement which ran until Michaelmas (29 September) 1884 would force Mr Scobell to pay whether it was used or not, it would seem unlikely that he or his estate would continue paying after that date if there was no associated income.

It is difficult to imagine that quarries in such a remote location with no easy access to the nearby railway and poor connections to the outside world could have been very profitable. Unlike in modern times, where we have machines capable of cutting, shaping and polishing granite, quarrying the stone in those days was pure hard work and needed large numbers of men to do the labouring. Granite is so hard that the only way it could be broken in those days, whether to remove individual stones from a large block or to remove a large block from the quarry, was to drill a series of holes into the rock using a sledgehammer and a sharpened chisel, with one man swinging the hammer and another holding the chisel and rotating it between blows. Removing a large block from the quarry meant drilling a line of holes, roughly 1½ inches in diameter and several feet deep, packing them tightly with gunpowder, lighting the fuse and running for your life! Splitting a large block into smaller ones would require a line of holes of roughly half an inch diameter, 4-6 inches apart and of similar depth, to be driven into the rock; into each hole, sets of what are known as feathers and wedges could be driven. The feathers are basically strips of hard steel which are pushed in either side of the hole, at right angles to the required cut line. The wedge is like a chisel which is put in between the feathers. The line of wedges was then repeatedly hammered in sequence until the rock split, hopefully along the right line! Many examples of rocks which have been split like this exist around the area, including some on the paving of the riverside path near Shaugh Bridge car park, between the footbridge and road bridge.

Industrial Archaeology of the Plym Valley

Nowadays, of course, we have machines which are capable of cutting thin slabs of granite for use as benches and worktops and polishing them to a glass-like finish; the men of the mid-nineteenth century, swinging hammers all day long, could not even have dreamed of such things.

The blasting of the rock would have provided a few large blocks which were capable of being worked, plus huge amounts of spoil; looking at the size of the spoil tips below each quarry and comparing them with the relative size of each quarry, it would appear that there cannot have been a huge volume of useful stone sold.

Dewerstone Cottage was the counting house for the quarries, with a smithy and stable attached and a hayloft upstairs. On the final demise of the operators, presumably around 1886, it reverted to the Goodameavy Estate and was let by the

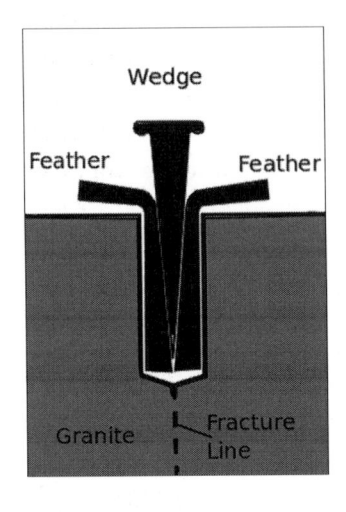

Diagram showing the use of a feather and wedge.

The spoil tip in front of the top quarry.

Dewerstone Quarries

owner to various tenants over the years. Nobody was shown in the census as living there while the quarries were in operation, so it must have been purely a working building. The earliest recorded tenants were Rebecca and William Gidley and their young niece Rosa, in the 1891 census. William was a general labourer; he and Rebecca ran a tea room there, as did many of the subsequent tenants. Mrs Mabel Legg, who lived there from 1932 until 1954, remembered that the blacksmith's shop was at the Goodameavy end of the cottage and when she and her husband moved there, the forge and chimney were still in position; they used to keep chickens there!

It would easily be possible to walk the area today and think there was little evidence of all the hard work put into these quarries; a closer look, however, reveals a wealth of information. Most people who visit the area do so from the car park near Shaugh Bridge, crossing the footbridge over the Plym. The granite-covered path which climbs up the hill from this footbridge is a modern improvement of one of the original access paths used by the workers. At the top of the hill the path turns left, and where it levels off this path joins the original lower level of the tramway which linked the quarries. Some of the stones on which you walk here are in fact the granite setts which supported the tramway rails; the holes for the locating spikes, and even an occasional spike, are still

Dewerstone Cottage in about 1910. (*NT*)

Industrial Archaeology of the Plym Valley

visible. Rounding the corner beside Pixie Rock (so called because it resembles a Pixie with a pointed hat!), you will see on your right a large gap in the rock face some 10 yards wide and 25 yards from the mouth to the far face, now much overgrown with trees. Look into the entrance, and on the left there is a rounded grassy area about 3 feet high surrounded by granite blocks. This was the lower quarry and the rounded area was where the loading crane was based. On the opposite side of the tramway is a spoil tip which, although not large, extends far down the steep side of the hill. The granite setts of the tramway peter out here due to the path having been 'improved' many years ago.

Continuing along the level tramway for a few hundred yards, you will come to a junction; the level continues but another track joins it from above. This is the remains of the inclined-plane tramway and here, again with careful examination, it is possible to see granite setts both on the inclined plane and also forming a junction with the level tramway. Walking up the inclined plane, it is possible in many places to see the original three lines of setts forming two tramways using a common rail in the middle. Judging by the marks on the setts, it looks as if the rails used were 5-yard lengths of the later type of pressed steel rail rather than the individual 1-yard, cast iron fish-belly rails as used on the early P&DR. The setts mostly have two fixing holes, with every fifth one on average having four holes where the rails joined.

Half way up was the passing section where the two tramways split, the central rail splitting into two separate ones so that the wagons could safely pass. The evidence for this is still there but much less visible than it was when I first visited the area 20 years ago. By my estimate, the distance from the start

Inclined-plane tracks.

of the junction to the passing section is about 135 yards; the passing section in the middle of the incline is about 55 yards long; then there is another 135-yard section to the top of the slope. On reaching the top, you come to the area where the cables ran back and forth, and between the rail setts there are other setts on which were rollers to support the cables over the apex of the hill. The area beyond is paved with granite and the trench where the lower cable came out from the bottom of the winding drum is covered with granite blocks. Granite, of course, was something of which they had plenty laying around so it was a cheap building material.

At the end of the paved area we find the winding house, which housed the large wooden drum. Cables ran around the drum and then down the hill, running on the rollers mentioned above. The axle and remains of the steel braking band are still there now, although the wooden drum has long ago rotted away. Beside the winding house I have found evidence of further setts leading into a short cutting in the hillside, probably the remains of a siding where the upper level tramway ran alongside the inclined plane. I have also found evidence of what might possibly have been a points system which would have allowed the wagons to be rolled straight onto the inclined plane, connected to the cable and lowered. This would have made enormous sense, of course; they certainly would not have wanted to unload and reload the stone at each junction.

Continuing along the upper level, you will first find on your right, about 80 yards along, a large piece of dressed granite, which presumably was being worked on the day that the quarries were closed down and was simply abandoned where it sat. Comparing this piece of stone, which had simply had its edges squared off using a hammer and chisel, with a modern, immaculately cut and polished worktop will show how rough the finish was and will give some idea of how much work with a hammer and chisel it would have taken.

Another 60 or so yards further on the right it is possible to make out below you, some 15–20 yards down the hillside, the remains of a building, probably a workshop of some kind. The walls of this building were shown complete but with no roof on the 1885 OS map.

Further on your left you will find the other two quarries; the first you reach is also the largest of the three, being 15 yards wide and 35 yards deep to the working face. The second, which still has the remnants of its crane base, similar to the one in the lower quarry, is the same width but not quite so deep. Both quarries have considerable spoil tips, built out like promontories from the hillside in front of them.

Industrial Archaeology of the Plym Valley

The inclined-plane winding house as it is now.

A piece of part-dressed stone.

From the end of this level a track leads uphill to the left, climbing through the woods to come out below Carrington Rock, the large granite outcrop, atop which can be found the following inscription:

CARRINGTON
OBIT
II Septembris
MDCCCXXX

This is a memorial to N. T. Carrington, the Dartmoor poet, who died aged fifty-three of consumption (tuberculosis) on that date (2 September 1830).

Dewerstone Quarries

Retracing your steps all the way down to the lower level and turning right brings you to the most visible reminder of the quarries' workings. This is Dewerstone Cottage, the old counting house, smithy and stable. The 1885 map shows this with no roof; possibly the Johnson Brothers left it derelict and the Scobells renovated it for occupation. Certainly it was occupied continuously, according to the census and personal reports, from 1891 up until 1954. Despite its remote situation and lack of running water and mains electricity, it appears to have been popular with tenants. It would appear that the cottage was used by most if not all of its tenants to provide drinks and snacks for the many walkers who visited the area. Certainly Bill Northmore, whose family lived there until about 1910, remembered his mother doing this. Frederick Ashcroft's family lived there between 1913 and 1930 and he recalled, as a child, walking to Goodameavy Farm to get milk, butter and cream for his mother to make cream teas.

Mabel and Charlie Legg lived there from 1932 until 1954 and they also provided teas for visitors, of which there were huge numbers in the pre-war days. Mabel recalled the trains back to Plymouth on a Sunday afternoon being so crowded that it was sometimes necessary to put on extra carriages and even open wagons to carry all the passengers. Charlie Legg built a waterwheel-powered electricity generator to provide power for the house, harnessing the power of the Blacklands brook, the stream which runs down by the side of the cottage. He also built one of the bridges across the Meavy which allowed the visitors easier access.

Despite the remoteness, Mabel Legg remembered her time there with fondness. Charlie rode a motorbike and she originally cycled everywhere until, during the

Dewerstone Cottage as it is now.

Industrial Archaeology of the Plym Valley

Charlie Legg's bridge over the Meavy with Shaugh Bridge in the background. (*NT*)

The tramway cutting between Dewerstone Cottage and the embankment.

Second World War, she also bought a motorbike to get to her job as a postwoman. Getting food supplies was difficult so they grew as much as possible themselves. By modern standards their life was very hard, but they loved living there.

In 1954 Colonel Hill, the much-loved owner of the estate and grand-nephew of Edwin Scobell, died; the Leggs left and the cottage became disused. With a sad inevitability, it became vandalised and was eventually burnt out. The National Trust acquired the estate in 1960 and leased the cottage to the Scouts, who rebuilt it to its present state, using it as a base for camping and outdoor activities up until spring 2009. At the time of writing it is very sad-looking, boarded up and disused.

Below Dewerstone Cottage are the remains of the Blowing House, simply a hollow dug in the ground roofed with granite blocks where the gunpowder for

Dewerstone Quarries

blasting had been stored. Little remains today apart from the long granite blocks which had covered it; the Scouts removed these and filled in the hole to stop people getting in there. The flat area beside the river below Dewerstone Cottage is where the Scouts used to camp.

The level on which the tramway ran continues past Dewerstone Cottage, past the small disused quarry on the right, then through a cutting and on to the start of the embankment, which was intended to carry the tramway across the river to join the railway. An arch bridge built into the embankment, recently restored by the National Trust, allows access to the alternative riverside path between the cottage and the road. From the start of the embankment, the current path veers right and continues along to join the road beside Joseph Scobell's bridge across the River Meavy. A stone set into the wall of the bridge shows the initials JS and the date 1832 to commemorate his building of the bridge.

Much of the embankment is still there, but many of the granite boulders from which it was constructed were removed and used to build Lopwell Dam in the 1950s. Although logic dictates that the tramway must have gone as far as possible to ease the transport of the granite, nobody can recall seeing granite setts or rails beyond the cottage. Mabel Legg was the last inhabitant of the original cottage, living there from 1932 to 1954, but her husband's family had lived in the cottage from 1910 to 1913 and she had never heard of there being any rails or setts there. Having said that, even by 1910 the quarries would have been disused for at least 20 years so I suppose this is hardly surprising.

Joseph Scobell's initials at Goodmeavy Bridge.

SHAUGH IRON MINES

Iron is a mineral which is not usually associated with Devon, but a large lode of iron ore existed in the Shaugh Bridge area. The earliest known mine in the area was the Shaugh Iron Mine, which was operated by Messrs Langdon and Paddon of Stonehouse from at least 1835, according to a lease dated Michaelmas of that year. The main working of this mine was on top of the hill some 500 yards south of Shaugh Bridge, on land which is part of the Maristow Estate. As this mine included much surface working, it is possible that iron ore had simply been dug from the ground for many years before this time.

The Mining Journal and Commercial Gazette of 8 July 1837 advertised the mine for sale, including the description, with perhaps more than a touch of hyperbole, of 'an immense lode of haematite iron ore intermixed with plumbago (black lead ore)' from which 'upwards of 20,000 tons may be raised annually with great facility and at a small expense'. As if all that was not sufficient inducement, it added that there were 'also indications of a copper lode, supposed to cross the lode of the Wheal Lopes copper mine in the adjoining sett' and that, 'The ores bear the high character with the ironmasters in Wales. The port of Plymouth possesses peculiar advantages as a place of shipment and the contiguous railways essentially reduce the expense of transit. In short, with a very modest capital, these mines, which are nearly inexhaustible, will be found to require only intelligence and attention to open a most prolific source of wealth.' For those who found themselves excited by such an unforgettable offer, the mine captain, Thomas Henwood, who lived nearby at Mount Pleasant, was available to show interested parties around. A. K. Hamilton Jenkin, in his *Mines of Devon*, suggests that the lack of interest in this glorious opportunity might have had something to do with the 'exorbitant dues of 2s 6d demanded by the mineral lord', at that time Sir Ralph Lopes.

At that time the mine appears to have consisted mainly of a long series of cuttings, starting by the side of the road from Shaugh Bridge to Nethershaugh and running north–south in Square's Wood, plus some other open works nearby. The advert states, 'From the localities of the mines, the works can be prosecuted without the assistance of machinery,' so it is possible no shafts had been dug at that

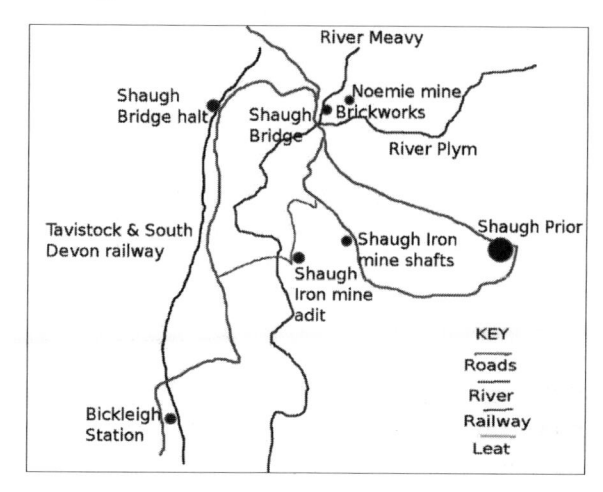

Shaugh Iron Mines

The Shaugh iron mines.

time. Further up the road was an old roadstone quarry which also contained some ironstone.

Perhaps surprisingly, there appear to have been no takers for the above offer because another advert appeared in the same publication on 31 March 1838, and again on 2 May 1840. By the time of the 1838 advert the mine had been developed considerably, because there was now mention of a shaft and an adit 'driven 90 fathoms in from the side of the hill, at a depth of 30 fathoms from the surface of the lode'; by 1840, a waterwheel had been erected which was fed from the River Cad (a.k.a. Plym) and this operated pumps and all the machinery. As the top of the shaft is at the top of a hill, the waterwheel must have been down near the adit portal by the river.

The 1840 advert also states that:

> Hitherto the ore has been carted to the Plymouth and Dartmoor Railway on Roborough Down, but Sir Ralph Lopes, having given permission for the making of a road through the level of Bickleigh Vale to communicate with the Cann Railway, (for the use of which, arrangements have been made with the Right Honourable the Earl of Morley), which forms a junction with the Plymouth & Dartmoor Railway, near Plym Bridge, a distance of about one half may be thereby saved, and the expense of transit to the waterside may be attained at a comparatively trifling expense.

The above demonstrates the difficulties of transporting materials at that time. The nearest junction with the Plymouth & Dartmoor Railway would probably have been at Jump (later Roborough), a couple of miles distant, or else through

Industrial Archaeology of the Plym Valley

the winding lanes via Leebeer to Tyrwhitt's Wharf above Clearbrook. Either of these options would have meant hauling carts along dirt tracks which ran up some hills steep enough to have tested the strength of the toughest horses. The new road down the valley to Cann to link up with the Cann Railway would certainly have made the total journey shorter and much more level, but still meant hauling heavily laden wagons roughly 3 miles over very muddy tracks which after heavy rain would have, and still do, become quagmires. For anyone to consider this an improvement shows just how difficult the alternative must have been!

Nothing is heard of the mine after 1840, so we must assume that it was not being worked. Possibly the booming price of iron ore around the 1870s is the reason why this mine came into production again. On 8 August 1868 Sir Massey Lopes, son of Sir Ralph, granted a lease for the mine to William Drew, described as a contractor, and William Godden, described as a mine agent. We know from other sources that Godden had been Captain of Boringdon Consols silver and zinc mine since at least 1852. Between then and 1854 he wrote a series of reports for the shareholders which were irrepressibly confident in tone, but which always seemed to preface a call on the shares, in other words a request for more cash. Given that Boringdon Mine was sold in 1856 and the company was wound up the following year, it comes as little surprise that Godden and Drew appear not to have been too successful at running this mine. It appears, judging by reports from Sir Massey's steward, that he made several calls to see Captain Godden, who was, apparently, never there!

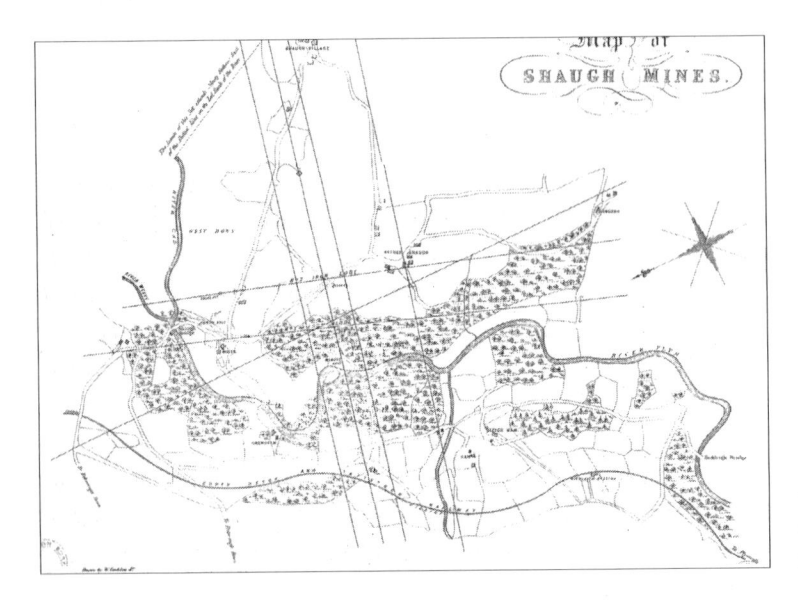

William Godden's map.

Godden does, however, provide us with a map of the area at that time. This map shows the mine shafts which had been dug down from the original surface workings, following the lode. The vertical shaft dropped from the southern end of the cutting some 230 feet (according to Dines's *The Metalliferous Mining Region of South-West England*) to the level of the river below, where it was drained by an adit at the foot of the precipitous hillside. The 1885 OS map also shows buildings near the adit portal. According to Maurice Dart in his book *Narrow Gauge Branch Lines – Devon Narrow Gauge*, a 2-foot gauge tramway issued forth from this adit and turned left down the track to the riverside as far as the bridge where the track crossed the river. Unfortunately, the accompanying photograph is very definitely not of this area, but if this was the case then the tramway would certainly have made life much easier for the miners; sadly, I have found no corroborating evidence that it existed.

From the bridge the ore could be loaded onto carts and taken up the track to the Shaugh Bridge–Bickleigh road and thence to a siding, perhaps half a mile distant, which had been specially built for this purpose on the Tavistock & South Devon Railway line above. This siding opened on 1 August 1870 and was eventually removed to make way for the Shaugh Bridge platform on 19 October 1907. While this system was in operation it meant the mine had the most efficient transport system available, the ore being taken directly from the working face to the railway as quickly and easily as possible. This would have been a vast improvement on the previously mentioned system.

From Lady Day 1870 a new lease was signed by Messrs John Brogden & Co., who according to the *Directory of Devonshire* had been operating the mine since

The remains of the bridge near Shaugh mine.

June 1869, further proof that Godden and Drew had not been very successful in their operations. This was the boom time for this mine. An article in *The Mining Journal and Commercial Gazette* gave the output of the mine in 1870 as 1,307 tons of brown haematite (iron ore) plus over 200 tons of mundic (iron pyrites), and in 1872 as 1,591 tons of haematite, and although production fell off after that date Dines gives a total output for 1870–4 as 4,670 tons. Dines gives no production figures after that date, but an article by Richard Meade, Assistant Keeper of Mining Record Museum of Practical Geology, gives a figure of 965 tons for the period 1874–5. The Brogdens were still paying rent for the mine up to Michaelmas 1880, although not afterwards. There are two points of interest here. Firstly, in view of Hamilton Jenkin's comments about dues payable to the mineral lord, their lease stipulated 9d per ton instead of 2s 6d, so this may have had some bearing on their success. Secondly, the peak output of 1,591 tons shows just how much wishful thinking was involved in the early advert, with its promise that, 'Upwards of 20,000 tons may be raised annually with great facility and at a small expense.'

A letter from one Thomas Gregory dated 13 July 1874 appears to relate to an inspection Mr Gregory did on behalf of Sir Massey Lopes. He suggests ways in which the mine could be operated, and his conclusion is, 'I have no doubt the lodes will be found of equal value which will yield a large and profitable output of ore for a long time to come.' Despite this optimistic outlook, it would appear that Sir Massey Lopes could not find anyone willing to carry on with this mine after this time.

John Brogden & Co. was the trading name of John Brogden and his sons Alexander, James and Henry. The Brogden family had originated in the Manchester area and had all become involved in the iron industry. John was born around the turn of the nineteenth century and by the 1870s, when he and his sons became involved in Shaugh Mine, they were all wealthy; the sons each had substantial properties in Wales. Another Brogden, George William Hargreaves Brogden, was about to become involved in the area as well; I can find no definite proof that he was related but he was also from Manchester, slightly older than John and also involved in the iron industry. Finally, one of John's sons was also a George W. H. Brogden. My feeling is that John and George were brothers.

In 1875 another mine was opened in the area. This was the Noémie Mine, just north of Shaugh Bridge, on land owned by Edwin Scobell of Goodameavy. The shaft of the mine was dug in the area between the rivers Cad (Plym) and Meavy, in an area which was no more than 30 feet above the rivers so must have been very prone to flooding. Water was diverted from further up the Cad to feed a waterwheel which must have been necessary to keep conditions underground even remotely bearable.

Shaugh Iron Mines

The Noémie mine was named after the French-born wife of one of the owners, Edward Casper. Casper had had an interesting life; born in Middlesex in 1824, by 1856 he was in Australia, presumably working in the mines there, and married to Noémie; his first daughter was born in Tasmania in that year and his second in Victoria the following year. In June 1875, in partnership with Frederick Lewis, he had taken out a lease from Edwin Scobell, who had in turn asked Sir Massey Lopes for permission to use (and return) water from the Cad to work the waterwheel, and also to build an 'ornamental bridge' to carry the iron ore to the railway siding at Bickleigh. Both these permissions were granted, and both Edwin Scobell and Casper and Lewis paid Sir Massey Lopes rents for the privilege.

The mine started producing iron ore in 1875, but the results were not good. It produced only 175 tons; in contrast the Shaugh mine, well past its peak performance, produced 965 that same year. The following year was slightly better, producing 222 tons, but this was nowhere near enough to be profitable. Some people have also suggested that possibly the ore was not of very good quality, and maybe because of this five tons were set aside in this year to experiment in brick-making. The following year the company name changed to Lewis & Co. and in December 1879 a company was formed called the Ferro-Ceramic Company Ltd; one of the directors of this company was the aforementioned George William Hargreaves Brogden. A new lease was signed between Edwin Scobell and Frederick Lewis on 1 March 1880, 'a lease of license and authority to mine for and dispose of the iron and iron ores under the piece of ground called Dewerstone Wood, in the parish of Meavy, in the County of Devon, and to work the brick and other clay under the said ground'. This is the first official mention of brick-making. Brick-making requires three basic ingredients: clay, sand and iron ore for the colour; all of these were readily available from the mine. Indeed, it may well have been the presence of too much sand and clay which reduced the quality of the iron ore, making it unsuitable for smelting. After building the kiln, the only major expenses would have been coal, labour and transport, so on the face of it this would seem to have been an ideal business to carry on here. In 1881 John Dester Canning and his wife Amy Prudence, both of whom were shareholders in this venture, were living at Clearbrook and he was listed as being 'Manager of brick and iron works employing 14 men and 1 boy', showing that at that time the brickworks appeared to be a success.

In 1880 a patent was filed for an 'Improved Brick Kiln', the patent holders being Messrs George Brogden and Edward Casper, both of whom were shareholders in the Ferro-Ceramic Co. In 1882 a further patent was granted for modifications to this patent. This improved brick kiln was what is known as a tunnel kiln, a long, narrow kiln open at each end with the furnace in the middle. There were four

Industrial Archaeology of the Plym Valley

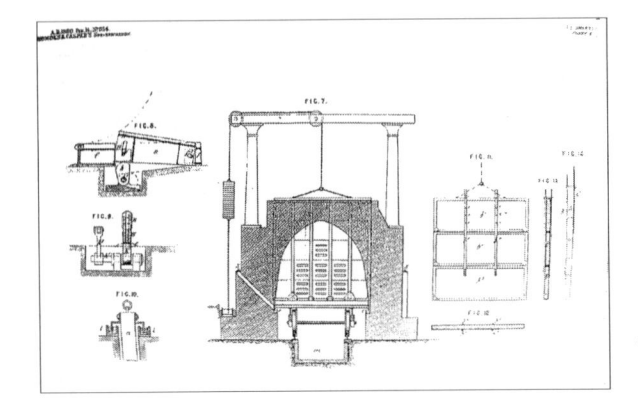

The brick kiln patent, showing a section of the kiln.

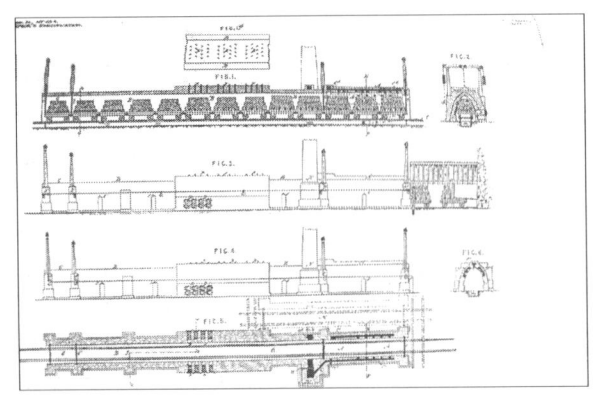

The brick kiln patent, showing the elevation and plan of the kiln.

sets of doors at intervals along the kiln, one at each end and two more inside the kiln on either side of the furnace. A set of rails ran the length of the kiln, running very gently downhill from the inlet end towards the exit, and the side walls had horizontal slots in them to allow the sides of the wagons to seal against the walls. This enabled soft bricks to be fed in at one end on wagons, the wagons rolling slowly down the slope until they encountered the first of a series of stops on the rails. These stops were controlled by levers from outside the kiln, allowing the wagons to roll on to the next stop after they had been in the kiln for the required time. The doors were counterbalanced so that they could be opened quickly to allow passage of the wagons, then closed to retain the heat.

After some time in the entrance section, where the bricks would warm slowly through, the wagons rolled on into the furnace section, where the bricks were fired to the correct temperature for the required time. The doors at each end of the furnace and the fact that the wagons sealed against the walls meant that the heat was concentrated where it was needed and was able to circulate among the bricks. After firing, the bricks rolled through into the final section, where they could cool slowly before emerging through the final door out into the air. The gentle warming and cooling made for more even firing of the bricks and less likelihood of them cracking, and the process allowed the kiln to be run 24 hours a day. This all comes across as being amazingly high-tech for the time.

After exiting the kiln, the bricks continued along the track, then via a turntable around to the storage area beside the kiln where they could be unloaded and prepared for transport. The 1885 OS map shows the whole of the kiln area under one large pitched roof. At the northern end there is a rectangular building which possibly housed a pugging mill; this was similar to a mincing machine and was used to mix the clay, iron ore, quartz and sand thoroughly and, more importantly, to make sure there was no air in the mix which would expand on heating and blow the bricks apart. This mill would also have been powered by the waterwheel.

As well as the brick kiln mentioned above, there was a smaller kiln for making tiles which appears to have been at the southern end of the brick kiln.

Despite the obvious advantages of this site and the available minerals, the business did not flourish. By 12 March 1883, following an extraordinary general meeting of the Ferro-Ceramic Company, the decision was taken, 'That it having been proved to the satisfaction of the Company that it cannot, by reason of its liabilities, continue its business, it is advisable to wind up the same voluntarily.'

The demise of the company may well have had a great deal to do with the existence of a brickworks nearby at Lee Moor, run by the Martin Brothers, who were the operators of Lee Moor clay pit. According to an article in the *Western Morning News* of 2 November 1864 this operation had four kilns and was capable of producing 100,000 bricks a month, all of which were made using the unwanted residues from the much larger china clay works there. This would have been a major and much more successful competitor even if their kilns were less efficient than the 'New Improved' one at Noémie. Their kilns had to be stacked with bricks arranged in a very complex way to allow the heat to circulate, then fired slowly at first, increasing the temperature towards the end to complete the firing, then allowed to cool before finally removing the bricks. The whole process took two weeks.

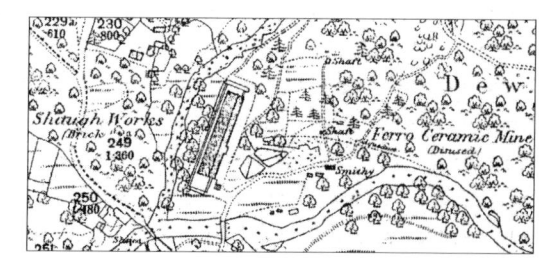

A section of Ordnance Survey map showing the brickworks and the Noémie mine.

Industrial Archaeology of the Plym Valley

Given that the Ferro-Ceramic company had nine shareholders, each hoping for some return on their not insubstantial investments, perhaps this was simply too small a mine to support their aspirations.

On 12 May of the same year, the lease for the mine and brickworks was offered for sale, including all plant, listed as:

> A Brogden and Caspar's improved tunnel kiln, tile kiln, 2 working sheds nearly completed adjoining the brick kiln, tramways, 18 iron tram wagons in good condition, 2 turntables, waterwheel 16 feet diameter by 36 inches, pump and bob with working and hauling gear, kibbles, bellows, anvil, vice and a large quantity of smiths and mining tools, wheelbarrows, winch, lot of tramway metals, carpenter's benches, lot of nails, timber, being materials for the construction of the proposed sheds and tram waggons, rolled iron etcetera., about 60 tons of iron ores and a large quantity of bricks and tiles, manager's office and smith's shop.

There is no record of anyone taking on the business, but Maristow Estate Rack Rentals show that Edwin Scobell continued paying annual rents – £20 for the use of the bridge and £30 for the use of water for the waterwheel – up to November 1884, large sums of money which would have been unlikely to have been expended without good reason. Although the 1885 OS map describes the mine as being disused, it shows the kiln still covered with the large roof and does not describe it as disused.

What, if anything, remains of all this enterprise today? The area of woodland around the Shaugh Iron Mine is still owned by the Maristow Estate, leased to the Forestry Commission, and is very definitely not accessible to the public. I was given permission to enter the area to see what is left, for which I am very grateful to Joe Hess of the Maristow Estate and Ben Phillips of the Forestry Commission. The answer is that very little remains, except some long, narrow and very deep cuttings at the top of the hill. These are well fenced off, as they should be; in places they are 30 feet deep and a fall into them could ruin your whole day. Near the river, the adit portal has been filled in and only a spoil tip near that adit and the track leading to the remains of the bridge across the Plym remain, apart from a tiny fragment of a wall where the main building was. The access tracks are also very wet, steep and muddy.

The Noémie mine and brickworks, on the other hand, are on National Trust land, very much open to the public, and there is much evidence of their existence. Directly in front of the footbridge which crosses the river from the car park is a pair of long narrow walls, obviously the ruins of a building. This was the Brogden and Casper Improved Brick Kiln. Although very much tumbledown, it is still

Shaugh Iron Mines

Above left: The large cutting at Shaugh mine.

Above right: The smaller cutting at Shaugh mine.

Right: The brick kiln and pugging mill. The horizontal slots for the wagon can be seen.

Below right: The brick kiln today.

A door slot in the brick kiln wall.

possible to make out some important details. The horizontal slots in the side walls where the sides of the wagons fitted are still visible, as are some of the vertical slots in which the doors ran. At the end closest to the bridge is a hillock which might have been the tile kiln mentioned in the sale notice, and at the other end is a rectangular building which probably contained the pugging mill.

Some yards east of the kiln are the remains of the building which housed the waterwheel. The water which fed the wheel ran from its source, which can be found further upriver, from which it would have been brought on wooden launders, no sign of which remain. It passed under the main track which leads up the hill via a conduit covered with large granite slabs which is still visible as a step in the path. It then flowed along a leat, which can be followed directly to the waterwheel house. The tailrace, which took the water away from the wheel, is visible as a channel curving back towards the river, which it meets slightly above the footbridge. In this area are also assorted depressions which appear to have been pits used for storing the various materials prior to use.

Walking up the main access track, you will see the remains of a building on the right; this was the smithy, and it is named as such on the 1885 OS map. A large tunnel on the left of the track opposite the smithy is the remains of one of the adits, the roof of which collapsed long ago, leaving a large depression behind the entrance. If you carry on walking from the adit entrance past that depression for 75 yards, you come to a large squared-off hole some 10 feet wide and 6 feet deep; this was the top of the main shaft, which was still open in the early twentieth century but was reputedly filled in by the farmer after a pig fell down it. This memory comes courtesy of Jim Williams, who was brought up nearby at Urgles Farm, and who went down the pit on a rope to retrieve the pig! Walking 40 yards south from the mine shaft, past a flat area where picnic fires appear to have been regularly lit, you reach the bank of the Meavy;

Shaugh Iron Mines

The adit beside the access track.

here you will find another tunnel, blocked by a steel grating and framed by the roots of a large oak tree; this was the other adit. Looking through the grating with the aid of a torch, it is still possible to see the adit running down for some distance.

Looking around here now, it is difficult to imagine that this mine once provided work, at its peak, for eight men underground and eight more on the surface.

Returning from the adit towards the current footbridge, have a look around on the ground. There are many broken bricks here on the main path, evidence of what was produced here 130 years ago. When you reach the bridge itself, note that it rests on a stone-built structure rising from the bed of the stream. This was originally one of the two supports for the 'ornamental bridge' built to carry the ore and bricks across to the road for onward transport; the second support was washed away long ago. Having had to pay wayleave for the bridge, it was demolished as soon as it was no longer needed, but the supports remained. For many years after the demise of the brickworks, Arthur Hill, nephew and heir of Edwin Scobell, refused to allow public access to the area, but when his son Colonel Gerald Hill inherited the estate in 1933 he was happy for people to walk there. At first, people would just jump from stone to

Industrial Archaeology of the Plym Valley

The collapsed mine shaft.

Above left: The waterwheel pit.

Above right: The adit by the riverside.

A view inside the riverside adit.

stone to gain access, a risky enough business for a man but even more so for ladies dressed in the long, heavy skirts of the day! Other bridges were later built here to improve public access, some across the Meavy, including one built by Charlie Legg who lived at Dewerstone Cottage and another by Fred Jeffery who lived at Kiln Cottage. The current one was built across the Plym in 2010 by the National Trust, replacing a previous one also built by them.

SHAUGH LAKE DRY

The National Trust car park beside Shaugh Bridge lies in the shadow of a substantial, stone-built sloping wall with gaps in it, reminiscent of some kind of fortification. Behind and beside this wall are other walls and buildings, now in a state of decay, but it is obvious that this was at some time a construction of some importance. What could it have been?

From about 1870 until 1960 this place would have been a hive of activity, because this was the site of the kiln which was used to dry the china clay from the Shaugh Lake claypits, part of the huge china clay extraction works near Cadover. Shaugh Lake is the northernmost of three claypits and the nearest to Cadover Bridge. Next to it is the much larger Lee Moor pit, and finally there is Headon pit. Shaugh Lake and Headon were owned by Watts Blake Bearne, a company which had been involved in the china clay industry locally since 1861, although the Watts family had owned clayworks since 1796. Lee Moor was owned in the nineteenth century by Martin Brothers, which became part of English China Clays and eventually Imerys.

Watts, Blake Bearne is still in operation, now known as Sibelco UK Ltd, and is an international supplier of minerals, still including china clay, and Shaugh Lake is still in operation. You will look in vain for the lake after which it was named; lake is a local name for a river or stream.

To explain the reason why this kiln and the nearby buildings and structures associated with it were constructed, we need to understand a little about the china clay production process. The china clay was (and still is) extracted in the claypits using water from high-pressure hoses which blast the soft clay away from the quarry face. The resultant mix of clay, sand, mica and water is known as slurry. Before it can be turned into usable china clay, this slurry first has to have the sand and mica and then as much as possible of the water removed before finally drying it out to form blocks which can easily be transported. I will follow this process in my explanation of the function of each building along the way, so we need to start at the top of the hill above Shaugh Bridge.

Shaugh Lake Dry

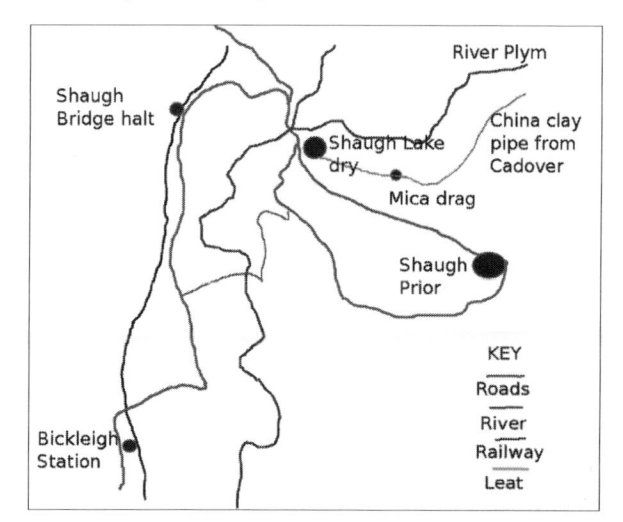

A map showing Shaugh Lake dry.

Shaugh Bridge car park today, showing the loading entrances.

Shaugh Bridge car park today, showing the kiln buildings above.

Industrial Archaeology of the Plym Valley

In the first stage of this process, the china clay slurry was piped down from the claypits along a six-inch-diameter pipe, parts of which are still visible, from Cadover Bridge through North Wood running parallel to the Plym. It flowed into a stone building, still visible if somewhat tumbledown and overgrown, at the top of the hill above Shaugh Bridge and across the river from the Dewerstone Rocks. This building was called a mica drag. Inside this building there were several broad wooden channels through which the slurry could be fed. Running the slurry through such a large area would reduce the flow rate, and lowering the flow rate allows the heavier particles, sand and mica, to fall out of suspension, while the lighter clay carries on with the water. Unusually, this drag was surrounded by a wall, which was there partly to keep out sheep but, more importantly, to protect the smooth flow of slurry from being disturbed by the strong winds which prevail in the area. The picture showing Melbur Micas in operation gives an idea of the process.

The remains of the supply pipe, the wooden launder which brought the slurry into the building and the wooden beams which supported the channels are all still visible inside this building, although very rotten, as would be expected after so many years. The building has also now been colonised by many young trees, so walking through it can be quite hard work.

The sand and mica residue would obviously build up and had to be cleaned out and dumped down the hillside once or twice a day. According to the China Clay History Society, this would have been done by diverting the flow of incoming slurry from each channel in turn, pushing the mica to a sluice area at the far end using a tool of the kind shown in the photograph, pulling out a plug of some kind and washing the mica away down through a drain. I can find no sign of this drain visible anywhere but the site is very overgrown and the drain would, of course, be below the working level. I have found the water supply though, a steel pipe with threaded steel connectors which the Institute of Plumbing tells me would have dated from the late nineteenth century. There would be no other source of water here, at the top of a hill, so it would have had to have been piped from further up the valley.

From the drain, it could be diverted to various dumping sites on the hillside below. It was initially dumped to the north of the drag, in what were called mica pits or dams, little more than holes dug into the hillside between the building and the river below, with the excavated earth piled up on the downhill edge to retain the mica and reduce, though not eliminate, the discharge of waste into the River Plym. These mica pits result in a flat area on the otherwise steep hillside, and judging by the prevalence of white waste in such areas, it seems that virtually any flat area on this hillside for about half a mile up-river

Shaugh Lake Dry

Melbur Micas, which would have been similar to the one at Shaugh. (*CCHS*)

from the kiln was a mica dam at some time. There is a huge flat area to the right of the track leading down from the mica drag towards Shaugh Bridge, overlooking the river and much used for camping, picnics and barbecues, which was a mica dam.

The 1905 OS map shows the mica drag still apparently complete and thus still operating at that time; it also shows the position of the mica pits down near the river. The results of all this dumping can still be seen today because the whole of the hillside, either on the surface or very close underneath, is white with clay waste. In later years, the mica was dumped on the hillside to the left of the path which runs from the mica drag down to Shaugh Bridge. Between that path and the neighbouring field, the long, narrow mounds are still there by the side of the track. One of these mounds shows up well in a photograph of the area taken, I would estimate, in the 1930s or 1940s. The mound is the long white area on the left.

Later, a more efficient mica drag was built at the Cadover end of the pipe, a concrete construction where 'paper alum', hydrated aluminium sulphate, was added to the slurry to speed the settling process. The slurry coming down would then have contained pure china clay which would have needed no more refining. A concrete reservoir was constructed further up the hill, in 1952

Industrial Archaeology of the Plym Valley

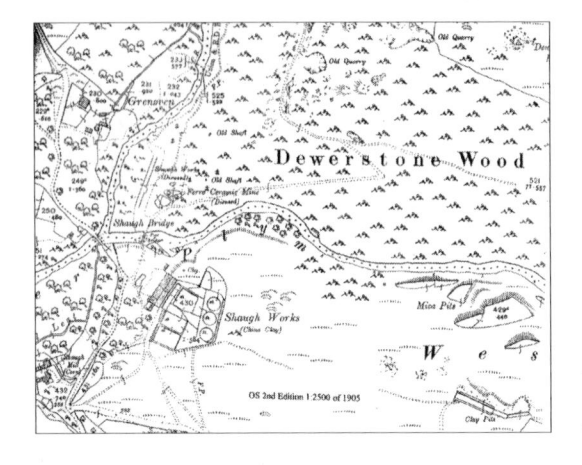

An Ordnance Survey map from 1905 of the Shaugh works, showing the mica pits.

The clay settling tanks and waste tips from Shaugh Beacon. (*NT*)

according to a date cut into the concrete, and a new pipe by-passed the old mica drag. We have to assume that this was the approximate date when the old mica drag ceased operating. Between here and the settling pits I can find no evidence of the pipe, which may have run deeper underground.

The slurry would then be piped down to the series of large tanks which are behind the kiln, first into one of the three round ones and then one of the three larger rectangular ones, all of which can be seen in the photographs above and below, where it was left to settle and thicken. In the picture below, the middle rectangular tank can be seen to be empty, so its contents would have been moved on to the next stage of the process. Finally, it was moved into the final thickening and storage tank immediately above the kiln and from there into the kiln as required.

The tanks were built in three stages, according to an archaeological evaluation commissioned by the Dartmoor National Park Authority from John R. Smith and the Royal Commission on the Historical Monuments of

England. The first stage, in the 1870s, was to build the three round tanks, the long narrow thickening and storage tank above the kiln and, of course, the kiln itself. The second stage, in the early 1880s, was to build two of the rectangular tanks, the ones nearest the road, and finally around 1895 the third rectangular tank was built.

These tanks are still visible now in the woods behind the kiln and are currently having some of the trees and vegetation removed, so they might be even more visible in future. The clay would settle to the bottom and the water could then be progressively drained off until the clay had thickened to the required consistency.

This thickened clay, by now resembling a thick, white custard, was pushed (it was too thick to flow very much by now!) through a system of sluices and conduits, still visible today between the settling tanks. China clay is thixotropic, which means that, like salad cream, it thickens up and can be thinned by movement. What this means is that it may be thick, but if you were to fall into it and start thrashing around it would get thinner and you could get stuck in it. Clearing out the bottom of the tanks thus needed great care.

It eventually poured (or was pushed) from the final narrow storage tank into the kiln itself, which in those days was completely covered by a huge pitched roof, the only part of the process which was under cover. The floor of the kiln was composed of fireproof tiles which were 18 inches wide. The thickness varied because near the furnace, where the heat was most intense, they needed to be thickest but the further away from the furnaces they were, the thinner they could be. These tiles rested on a series of underfloor walls 18 inches apart which ran the length of the kiln, forming a series of flues. There were twelve flues, so the floor of the kiln was 18 feet wide. The flues were fed with hot air from one or all three of the coal-fired furnaces which were at the right-hand end, nearest the road, of the ruins you see now. The furnace entrances are visible since the vegetation was cleared, and inside the innermost furnace the grate and remains of some sections of the flue walls can still be seen. The hot air from these furnaces ran under the length of the kiln and then fed into a chimney stack, the base of which was at the northern end of the ruins. The stack was blown up by the Royal Marines as an exercise after the kiln fell into disuse, to prevent it becoming unsafe, but the recent clearance work has uncovered the exit flue and the base of the stack beyond.

These furnaces would have needed a constant supply of coal to keep them fired. Richard Walke, whose family was at Leebeer Farm, recalls his father taking the clay blocks up to Bickleigh station with a horse and cart, then returning with coal for the furnaces. The road from the bridge to Shaugh Prior today is a steep, straight tarmac one, but the old road used to go down past

Industrial Archaeology of the Plym Valley

A close up of the clay settling tanks. (*NT*)

The kiln and cottage, showing loading bays with trucks loading. (*NT*)

Shaugh Mill, then up the valley to the village, appearing near the White Thorn pub. To get to the furnaces, they had to take a sharp turn left off this road beside the Mill and up over a steep slope which is now the drive for the house called Endomoor. They then went along the track, still visible today, which leads from the current road towards the furnaces. This was known as the coal road.

The firebricks and tiles which made up the flues would break up if they were constantly cooled and heated, so as much as possible the furnaces were kept fired 24 hours a day. As soon as the slurry came into contact with the hot floor of this kiln it would start drying out, producing clouds of steam in the process. David Fookes, who grew up at Shaugh Bridge in the 1950s, told me that the local children used to play a game, creeping in through a rear door near the chimney and seeing how far they could get through the steam before the workmen saw them! The kiln would have been running night and day, drying the china clay slurry and turning it into solid blocks ready for transport to the industries which needed this material.

Why was the kiln built here, in the middle of nowhere? Why not at Cadover, where the rest of the works was? The answer to that question is transport and access to the nearby railway. Initially, the dried clay was transported from the kiln by horse and cart up the steep hill to Bickleigh station, a tough haul for a pair of horses. Later it was taken to the much nearer Shaugh Halt and, with the advent of motorised transport, it was taken by lorry. Whichever station it arrived at, it was then loaded onto railway wagons and could easily be transported anywhere in the country. All this piping and hauling appears a bit long-winded but it was much easier in the late nineteenth or early twentieth

The inner furnace, showing the grate and flue walls.

century to do this rather than shipping it directly from the quarry at Cadover. Gravity powered the initial movement of the slurry, so it cost nothing to run, and the roads near Shaugh Bridge, although primitive at the time, were much better than those around Cadover. Martin Brothers, who worked the neighbouring Lee Moor pit, initially dried their clay there and carried it to Plympton by pack horse; later, they piped their clay slurry all the way to their kiln at Marsh Mills.

When we think of china clay we immediately think of its role in the manufacture of porcelain, but it has other uses as well. It is an ingredient in various products of the pharmaceutical industry and in the making of good-quality paper. China clay by itself is much too brittle to use; in order to make porcelain it has to be mixed with ball clay in roughly equal proportions. Ball clay is the much darker clay used in rough pottery such as terracotta (which means literally 'cooked earth').

When the kiln was in operation, it would have employed four or five people: a fireman to look after the furnace, two or three men to control the movement of the clay and probably an overseer or foreman. One of the men would probably also have doubled as the lorry driver – in those days there would be no employees standing around with nothing to do!

Imagine the scene inside the kiln. With the furnaces running and the steam produced by the drying clay, the conditions inside the roofed area would have

Shaugh Halt in the early twentieth century. Note the lack of trees! (*NT*)

been hot and humid, like working in the Tropics. Of course, much of the work was outdoors and when they went outside they were back in the cold, damp weather of England; the constant changes meant it was not a particularly healthy environment to work in. Moving the clay into the kiln was hard work; they pushed it along the conduits with a 'shiver' (pronounced sh-eye-ver), an implement similar to a large hoe with a metal plate on the end of a wooden stick, as can be seen in the Melbur Micas picture. Once the clay in the kiln had reached a depth of about 9 inches it was smoothed over using a planer, basically a long plank; the surface was marked out into 9 inch squares and then left overnight. This would be enough time to evaporate the water from the clay, but not long enough to fire the clay and turn it into china.

As the clay dried out, the squares marked on the surface would crack through, forming blocks. The following day the men would jump down into the hot kiln with a shovel and throw these blocks through the gap between the kiln wall and the roof. The floor of the kiln was so hot that the workmen wore wooden clogs to walk on it while removing the clay blocks, and even these clogs needed to be doused in water from time to time when they became too hot! Imagine the dust! China clay is a very fine powder, similar to talc, and when they tried to move the dried clay the dust would fill the air, already stifling from the heat of the furnace. They would finish work looking like snowmen, but the dust wiped off their clothes very easily, I'm told.

Obviously the temperature of the kiln floor would vary along the length, so the amount of time taken to dry the clay varied along the length as well. The area nearest the furnace would be cleared and refilled daily, while the area furthest away might be cleared less frequently, only once a week perhaps at the furthest end.

The blocks dropped 15 feet onto the linhay (pronounced 'linney'), which is the storage area immediately behind the sloping wall beside the car park. The gaps on

Wheal Martin clay dry, a similar layout to Shaugh. (*CCHS*)

the walls were to allow carts or lorries to reverse in for loading and the linhay floor is built up to a level which would have allowed easy access for loading. Here, the china clay was stored until it was needed to be transported for final use. If more space was needed, they would put planks across the loading bays and fill the whole area. A picture shows Horace Jeffery standing beside his lorry outside the linhay and in the background it is possible to see the blocks stacked up as described.

The five gaps in the wall were wide enough to allow the lorries of the day (at one time the lorries used were made by a firm called Peerless) to reverse inside and load the dried clay, although they are only just big enough for a modern car, and much too small for modern lorries! If the bays had been planked over, the clay would have been loaded from the entrance, then the planks gradually removed and the lorry reversed inside. The 15 feet drop from the kiln would smash many of the blocks, but this didn't matter as they would need to be crushed to powder before being turned into china.

For the men working here this was very hard work, much of it carried out in a hot, dusty environment and with the constant danger of falling into the tanks or kiln, all of which would no doubt give the Health and Safety Executive apoplexy today! Nevertheless, according to those surviving ex-workers who were interviewed, the job was popular with the men who worked there.

This site has one more surprise. At the right-hand end of the sloping wall beside the car park, beneath the furnace area and covered by the same huge roof which covered the kiln, there used to be a cottage. The cottage appears to have been built at about the same time as the kiln, certainly by the early 1880s, because there is no difference between the outline shown on the 1886 and 1905 OS maps, and it was definitely there and occupied in 1905. It was very small, long and narrow with few windows and only parts of some walls are visible now. The cottage was inhabited by the families of men who worked at the kiln, including Mr Jeffery and Mr Pundsack. Purely by chance, when I was helping with the clearing operations there, Mr Brian Pundsack arrived to see what was happening. He had lived in the cottage until he

Horace Jefferies outside the loading bays. China clay blocks can be seen on planks in the entrances. (*NT*)

Shaugh Lake Dry

The loading bay entrance.

was five years old, when the family moved to Wotter. He described it as being cold, damp and dark, with no electricity or running water; in comparison, the council house in Wotter where they subsequently moved, with lights which came on at the touch of a switch, plus an internal toilet which flushed at the pull of a chain, was unimaginable luxury.

The workers who occupied the cottage would have had to oversee the loading and weighing of the lorries, and for that there was a weighbridge. The original weighbridge was on the other side of the road, close to Shaugh Bridge itself, and the square office which housed the weighing equipment is still there, covered in a dome of concrete and incorporated into the roadside wall. Later, as the lorries became larger and would no longer fit onto the weighbridge, which had originally been built for a horse cart, a larger weighbridge was installed in front of the cottage. The remains of this can still be seen; the outline of the concrete surround is visible in front of the remains of the cottage.

Among the earliest families who lived in the cottage was that of Mr Fred Jeffery. Mr Horace Jeffery, son of Fred, was interviewed about his memories in 1987 and he said he thought they had lived there from around 1885, soon after it was built, and certainly their names appear there in the 1901 census. He recalled his mother's meat safe being near the back door of the cottage, and the wash house and meat safe are still visible today, as is the remains of the outside toilet.

Mr Pundsack's father, Toby, lived in the cottage from 1948 until 1952, and when he left a Mr Peters lived there until about 1954, when it was left uninhabited and allowed to decay. The clay kiln ceased operation in 1960.

Apart from the cottage and the linhay walls, which are visible from the car park, the kiln and settling tanks are in quite good condition, if overgrown with vegetation, and much can be seen. Behind and above the linhay walls is the kiln itself, the floor of which is very broken up, and the kiln furnaces are visible at the right-hand end. This can all be seen to the left of the slightly slippery and uneven steps at the road

Industrial Archaeology of the Plym Valley

end of the car park, which lead to a path. Behind the kiln are the large rectangular settling tanks; although much overgrown by young trees, their general features can easily be made out, as can the sluices and conduits through which the clay was moved.

These conduits are up to three feet deep, so while it is possible to inspect the site much care needs to be taken where it is overgrown. As you continue up the path, the three smaller, round settling tanks are visible just before the boundary fence. There are even the rotted remains of a wooden launder which would have been used to distribute the clay beside one of these tanks.

Just past the round settling tanks, a stile leads onto a path which continues up the hillside, and off to the left as you ascend you can see the large, flat area I have already mentioned which was a mica dam. The soil here is pure white under any top covering and burrowing animals and insects throw up spoil heaps which are as white as snow. Further up the hill, the ruined walls on the right are the remains of the mica drag, with a long, narrow mound below it which is another mica dump. Near the very top of the slope on the left of the path is the later concrete reservoir, from where the pipe track, a popular walk, continues through North Wood and on to Cadover Bridge.

The pipe which used to carry the clay slurry was a 6-inch internal diameter earthenware pipe, similar to a soil pipe which takes the sewage away from a house. Near Cadover, where the pipe crosses a small valley and a stream, there is a cast iron section, presumably to cope with the extra pressures caused by the drop into the valley. Thereafter it reverts to earthenware and the remains of this pipe are easy to pick out for much of the length of the track despite the efforts of various idiots to smash it. In places the pipe was doubled up, presumably to improve flow rates, and periodically there are inspection chambers to allow for clearance of obstructions, etc. Through North Wood they are roughly a quarter of a mile apart, but near the top of the slope they are closer, sometimes as little as a hundred yards apart.

Far left: Kiln Cottage in the winter. (*NT*)

Left: A wooden launder beside the upper settling tanks.

SHAUGH MILL

Shaugh Mill is located just downstream from Shaugh Bridge, beside the old road which ran down from Shaugh Bridge then up through the valley to the village of Shaugh Prior on the hill above. Coming from the Bickleigh direction, the old road turns down to the right after passing over the bridge, whereas the current road continues straight up the hill. The mill is on the right at the bottom of the hill, where the road splits. The old road for Shaugh Prior turns sharply to the left, with another road leading straight on to the hamlet of Nethershaugh. This area, like the woods above where the Shaugh Iron mine is located, was originally part of the Maristow estate.

The leat which fed the mill's waterwheel is very short when compared to other local leats, running from just below the bridge, a distance of little more than a hundred yards in a not very steeply sloping area of the valley. With no dam or weir to ensure a head of water, it is easy to imagine that in summer maintaining the necessary steady flow would have been very difficult. However, they obviously managed very well for something like 150 years, although it must have needed constant attention because the entrance to the leat is now totally silted up.

The first references to Shaugh Mill are in the local eighteenth-century newspapers, including the *Exeter Flying Post*, where it is described as a paper mill. During the eighteenth century there was a rapid expansion of paper making, at that time a process which made individual sheets by hand and was thus very labour-intensive. This continued until the early nineteenth century, when paper-making machines were invented, producing continuous rolls of paper of consistent quality much more cheaply than would be possible by hand. This was the reason many mills, this one included, changed use at that time.

The paper-making process in the eighteenth century involved collecting old rags, specifically cotton and linen, which were boiled with wood ash, which acted as a bleach, then washed to remove impurities and pounded to form a suspension of individual fibres in water, which I have seen described as resembling a batter. A mould, which was a rectangular wooden frame with a fine mesh base, was dipped into this suspension and swirled to ensure an even coating of the fibres on the mesh;

Industrial Archaeology of the Plym Valley

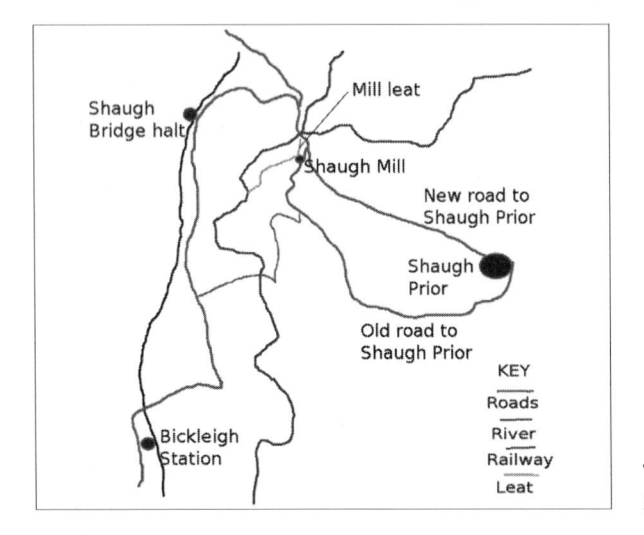

The location of Shaugh Mill.

Shaugh Mill in the early twentieth century. (*NT*)

then another tight-fitting mould was pressed into the first to squeeze out some of the water, leaving a thin sheet of fibres. This sheet was removed from the mould and piled up with other sheets to be pressed between layers of felt to remove more water.

Once the sheets had dried sufficiently, they were individually dipped into a tub of sizing solution, which varied in content but was generally a starch of some kind. This was done partly to waterproof the paper, which otherwise would act much like modern kitchen rolls, soaking up the ink to form blotches and making writing impossible. Obviously the whole process described here was highly skilled and very labour-intensive and a small mill such as Shaugh might have produced only a few hundred sheets of very expensive paper per day.

Shaugh Mill

The silted up mill leat.

The purpose of the mills in this process was to pound the rags, smashing them up into their constituent fibres. They used the waterwheel to drive these pounding mills, which were very similar to the clash mills used to refine tin ore. The shaft from the waterwheel lifted steel-tipped wooden shafts which then dropped onto the cloth, smashing it apart. Paper mills thus needed a good supply of clean water, both to turn the machinery and for the washing of the paper.

Obviously the conversion of a paper-mill to a grist mill, i.e. a mill for grinding grain, was a major undertaking, requiring completely different machinery for the totally different process. The pounding mills would be removed and replaced with grinding wheels, plus all the associated equipment required to receive and handle both the grain and the resulting flour. Using this equipment required vastly different skills which could not be learned overnight, so it would also possibly need a change of owner and certainly a change of operator as well.

We have some references to Shaugh Mill during the eighteenth century, mainly thanks to the Exeter Working Papers. In 1772 the miller and paper maker was one William Perry and his name appeared in the newspapers when two apprentices, William Saunders and George Starling, absconded from Shaugh. Perhaps the mill was too remote for them, or maybe he was just not a very good boss!

The mill was badly damaged by fire in 1788. Richard Howard was bankrupted here in 1790 and the mill sold in 1791. The owner and paper maker from 1803 until certainly 1816 was Mr Johnson Atkinson but the change of use from paper to grist (grain) mill appears to have happened around then because on 29 June 1818, we find the following advert placed in the *Sherborne and Yeovil Mercury* offering the lease for sale. This would of, course, fit in with the aforementioned arrival of paper-making machines and a consequent fall in the price of paper.

Shaugh and Bickleigh Mills in Devon

To be sold with immediate possession for the remainder of a term of 99 years determinable on the death of one excellent life subject to two annuities, amounting to 16 pounds a year under such conditions as will be produced at the time of sale. All these capital and well accustomed water grist mills which are supplied with an abundant and never failing stream of water called Shaugh and Bickleigh Mills, together with a dwelling house and all suitable outhouses, 4 young prime orchards now filled with great abundance of remarkably fine fruit, 3 meadows, a small plot of barley and nearly 7 acres of coppice, the whole forming most compact and desirous premises and consisting of up to 30 acres of ground with an unlimited right of common and turbary [this was the right to cut turf or peat from an area of bog] on West Down and Ham Green, which said premises are situated near Shaugh Bridge in the parish of Shaugh, distant about 3 miles from Plymouth and 6 from Tavistock. Also to be sold with the mills the benefit and advantage of 2 policies of insurance affected on the life thereon to the amount of 800 pounds. For which purpose a public survey will be held at the house of Mr. Congdon, known by the sign of the Commercial Inn, in Plymouth, on Thursday the 9th date of July next precisely at 12 o'Clock in the forenoon. In the meantime the mills may be viewed by applying to Mr Peter Holmes therein or to Mr William Atwill who lives in Bickleigh nearly opposite to the said mills and any further particulars may be known of the office of Mr Tolcher, solicitor in Plymouth.

NB the purchaser will be required to take the present bearing of apples at a fair evaluation.

They certainly knew how to write adverts in those days! Even the apples had to be accounted for.

We have no information as to who, if anyone, took on that lease in 1818, but in the 1841 census the miller was shown as George Luscombe, with his wife Elizabeth, also 25, and their children George, 2, and Fanny, 1. Also living there was John Tucker, 15, presumably a labourer working for them.

Between 1851 and 1866 John Mumford was the miller. The 1851 census shows him to be 22 and unmarried. His brother Edward, 18, worked there as a 'journeyman miller' and his cousin Sarah, also 18, was the house servant. They even had a waggoner, Thomas Williams, 22, living there as well, so we have to

Above left: The original Shaugh Mill waterwheel. (NT)

Above right: Horace Jeffery's waterwheel and generator. (NT)

presume that business was good. John Mumford was still there in 1861, but by then he was married to Sarah Maria, 24; they had two sons, Uriah Edward, 3, and John James, 1 month, and a daughter, Mary Margaret, 2. Of his 'journeyman miller' brother there is no mention.

Mumford had left by 28 August 1866, because the *Western Daily Mercury* carried the following advert.

> To Millers
>
> Flour and Grist Mills
>
> To let. Shaugh Mills (Adjoining Shaugh Bridge) with new and most improved machinery, and constant supply of water for working two pairs of stones. A dwelling house and all necessary outbuildings with about 8 acres of land. One mile from Bickleigh station. Immediate possession may be obtained - apply to Mr. Moon, Maristow, or Mr. C.L. Radcliffe, Plymouth.

The 1870 Directory of Devonshire names John Tregillis as the miller, although he was presumably the owner as the 1871 census says that Thomas Anning, 40, was the miller, with his wife Emma, 44, and their sons Charles, 15 and William, 13. John Barker, 26, was also listed there as a miller and Mary Beer, 14, was their domestic servant.

By the time *The 1878 Directory of Devonshire* appeared, William Harris, 44, was the miller at Shaugh Mills, and according to the census he was there until at least 1891. He was a Cornishman, born at St Pinnock near Dobwalls, and had only recently moved to the area with his family, in fact within the previous five years judging by the fact that all his children were born in Cornwall except the last. According to the 1881 census his wife was Marria (sic), 43. Their son William was 17, and their daughters were

Marria, 14, Laura, 12, Bessie, 5, and Mary, 3, who was the only one born at Shaugh.

Harris appears to have had a fairly shaky start with the business, because the Maristow Rack Rentals record shows him being behind with his rent in 1880 and early 1881, but by Michaelmas 1881 he had caught up and from then on paid on time. In 1891 the family is still there, with none of the daughters married. William junior is also a miller at that time and is shown as married, but with no wife in evidence.

From about 1895, according to Horace Jeffery who lived across the road at Kiln Cottage, the Edwards family were at the mill. Father Joseph, 67, is listed as living there but shown as a farmer in the 1901 census; his wife Rebecca was 64. Their eldest son James, 29, is described as 'farmer's son, miller', their daughter Florence is 27, and their sons Walter, 23, and Ernest, 20, are described at that stage as 'farmer's sons'. They had a visitor, John Worth, 29 and single, who was living on his own means. James and Walter jointly ran the mill until at least 1919 while also farming Mount Clog farm, which is at the top of the lane which branches left off the modern road to Shaugh Prior, roughly a half a mile from the mill. James is listed alone as miller and farmer of Mount Clog in 1926.

Mr Jeffery also remembered Mr Edwards using the mill wheel to turn a dynamo which generated electricity for both the mill and Kiln Cottage. By that time the wheel was in a bad state of repair and really needed to be completely replaced, which is why they gave up milling. Mr Jeffery said, 'He had a dynamo there which provided light for himself and the kiln cottage as well. Unfortunately the wheel was heavier one side than the other. As the wheel went over one side the light was bright and on the other side it was dull.'

When he left the mill in about 1930, Mr Edwards retired to Mount Clog farm and Mr R. D. Smale Willis (known as Dickie) lived there. He had been head gamekeeper for Sir Henry Lopes of Maristow, later Lord Roborough, and previously lived in the cottage at Bickleigh Bridge. They made teas for the day-trippers at the mill and apparently did well at it. Mr Jeffery and his father made a small water wheel away from the mill, diverted the leat and installed it as an overshot wheel. This wheel was again used to supply electricity and was much more successful than the old one.

Mr Willis lived there until after the war; Mr Toby Pundsack remembered him there in 1948 but by then there was no wheel working. The mill is now simply a private house.

RIVERFORD

Upriver from the weir beneath Cann Viaduct, the River Plym runs in more or less a straight line, with the path running parallel to it along the western bank. This is a lovely quiet stretch of the river where I have seen kingfishers, herons and sometimes even a cormorant fishing; brown trout plus the occasional salmon are to be seen in the river. About a half mile north of Cann the path skirts around a hillock, then continues straight on again beside a large meadow, which in spring is filled with bluebells. This meadow looks incongruous today in an area where generally both sides of the valley are steeply wooded hillsides coming right down to the river bank. This is an area which has fascinated me ever since I was told that it was the site of a hamlet, including a large farm house, and shown pictures to prove it. This is Riverford, the hamlet which disappeared! Few people looking at the site today would ever guess that here was once a thriving community with a working slate quarry and farm, home to as many as thirty-five people at its peak in 1881 (although that must have been a bit of a squeeze!), and that some of the houses were inhabited up until the mid-1950s.

The area which I know as Riverford, lying between the aforementioned hillock and Riverford Viaduct, and stretching from the riverbank up to the Plymouth & Dartmoor Railway line high above, had a very chequered ownership. In the early nineteenth century, when it was first inhabited, Riverford was almost surrounded by the Colwill Estate, at that time owned by major landowner and Lord of the Manor of Eggbuckland, Christopher Tolcher. Immediately to the north was a small area of land belonging to the Lord of the Manor of Efford, Erving Clarke, with the huge Maristow Estate belonging to the Lopes family, now Barons Roborough, to the north and east of that. Riverford was owned in 1840 by one Thomas Briggs, who lived nearby at Fursdon, half way down Blunts Lane and not far from the present-day Derriford Hospital. The 1840 tithe map of Eggbuckland shows just one building at Riverford, surrounded by a garden wall and an adjoining meadow, all of which was leased to one Thomas Martin. This building was the one I know as Riverford House, the largest building there, situated beside

Industrial Archaeology of the Plym Valley

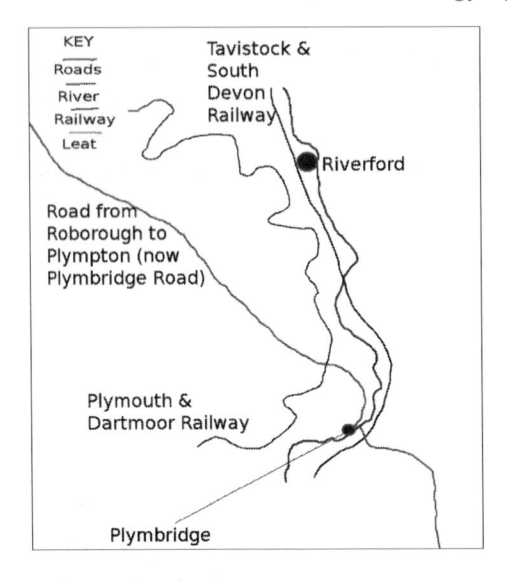

The location of Riverford.

Riverford hamlet in the early twentieth century.

the meadow. At that time there were no other buildings on the whole of the hillside and only farmland above.

No direct access track was shown on the tithe map, access being down the track which still exists, running today from Woolwell roundabout parallel to the stream which joins the river a quarter-mile further north. This path runs under the later-constructed Riverford Viaduct, originally built between 1853 and 1859. Neither was the adjacent Hyren Quarry shown, so we must assume it was not operating at that time. However, in the 1841 census we find that living at Riverford House (then called Allters Ford for reasons unknown) was John Martin, 50, and his family. John gave his occupation as slate quarryman, which presumably means that Hyren Quarry was in the process of being opened up.

If Hyren was not operating in 1841 it must have opened soon after, and a huge amount of work was done there. The hillock around which the riverside path skirts is a spoil tip, made up of waste from the quarry. The 1853 map showing the proposed route of the Tavistock & South Devon Railway shows the line of an inclined-plane railway from the quarry directly up the hillside to the Plymouth & Dartmoor Railway above; the zig-zag track known as Stuckey or Stuggy Lane also led up to that railway and another track went straight up the hillside to the railway track and on to the main road above, emerging in the area now known as Forresters Business Park. Logically, at least one of these access tracks must have been built as soon as the quarry was working, and we know the inclined-plane leading to the railway track, which was steam powered, had been built by 4 September 1847 because we find an advert in the *Plymouth, Devonport and Stonehouse Herald* on that date advertising an auction which included the following:

> Lot one will consist of all that superior steam engine 15 horse power with boiler fixed complete with copper piping ditto now erected on the above quarry together with carts and other apparatus thereto belonging. The steam engine is quite new and has been recently erected on the quarry and is of the best material. In addition to its being fitted for raising the slate from the quarry, it has all the requisites for being used as a saw mill, having a circular saw bench complete with machinery.

Also offered for sale were 'a rick of hay, a fine chestnut colt with gig and farm carts'. Henry Crace, tenant and occupier thereof, must have been living there, farming and running the quarry as well. He appears to have been made bankrupt as the person organising the auction is a bailiff.

On 11 May 1848, *Trewman's Exeter Flying Post* carried the following advert:

> To be let for Tender for 7, 11, or 14 Years, from Lady-day
> last, RIVERFORD, with its neat Cottage and Outbuildings and HYRON SLATE QUARRY
> lately occupied and worked by Mr. Crace, in Egg Buckland, Devon, containing altogether about 20 acres of excellent land, through which the Plymouth and Dartmoor Railway passes, and distant about 5 miles from Plymouth.
>
> The conditions under which the same will be let may be known of Mr Vallack, Solicitor, Torrington, by whom tenders will be received until Saturday the 20th day of May instant.
> 1st May 1848.

Although he practised at Torrington, Henry Adoniah Vallack was a local man, born in Kingsand, just over the River Tamar in Cornwall, in 1807. The 1853 deposited plan for the Tavistock & South Devon Railway lists all the landowners along the proposed route, and it shows that at that time he and his wife Elizabeth were the owners of the land and joint occupiers of the quarry along with one Francis Sellick. Sellick was described in the 1851 census as 'Farmer of 15 acres, Tea House'; he was then living at Riverford House along with his wife Harriott, 50, their son George, 19, described as 'working at slate quarry', and the rest of their family. They were the first people to start the long tradition of providing teas for visitors to the area, although it's difficult to imagine that there would have been many visitors to that remote, industrialised area surrounded by farmland in an era when few people had much leisure time.

Also living in the house were John Moses, a 27-year-old cordwainer (what we would now call a shoemaker), who was married to the Sellicks' eldest daughter Harriott, and their family. The whole Sellick family were local, all being born at Shaugh Prior.

This 1853 map shows for the first time that there were now two dwellings at Riverford; apart from the original house, there was now a slightly smaller cottage just up the track and facing at 90° to the house. Living at Riverford Cottage were Richard Creber, 54, described as a cart man, his wife Elizabeth and their family. Living in such a remote place, with no competition, he presumably found plenty of work, possibly carrying some of the slate from Hyren to Plymouth for sale or shipment and also carrying goods and produce from the farm and bringing in everything necessary to both farm and quarry.

Further up the hill, near the Plymouth & Dartmoor Railway track, there were also three cottages which appear to have been built around 1840. They are collectively described in census returns as 'Common Wood', but according to the 1865 OS map the only building shown is on the present site of Common Wood Cottage. Although it is outside of my area of interest, I will make mention of them because although many of the inhabitants worked on the local farms, some were involved in the

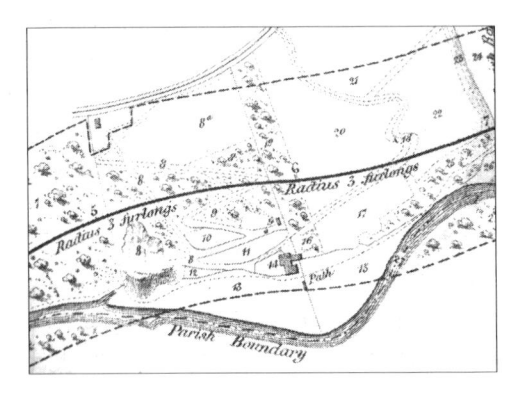

A plan from 1853 showing the proposed route of the Tavistock & South Devon Railway. The Plymouth & Dartmoor Railway can be seen at the top.

Riverford

Riverford Cottage.

quarries. The 1841 census shows just one family, that of a mason called Thomas Trenamen, living there but by 1851 there were five families there, all farmers except Thomas Rice, who was a 40-year-old slate quarry worker, and his son Richard, 10, an apprentice blacksmith who may also have been involved in the quarry.

By 1861 all the inhabitants of Common Wood were involved in farming. Down at Riverford, Francis Sellick and his family were still living in Riverford House, but according to the census return it is now called Ford Farm. He describes himself as a farmer of 15 acres again but the Tea House has now become a Cider House. This must be the origin of the story told by Mrs Joyce Fry, who lived there in the twentieth century, that Riverford House was once a pub called the Owl Arms. This, I would imagine, was much more likely to attract custom than the previous Tea House, especially among the thirsty quarrymen and miners locally.

The Sellicks' son James was 27 and a blacksmith and lodging with them was James Smallbridge, 54, described as a cleaver in a slate quarry, which might mean that Hyren was still being worked by Sellick at that time although I think it very unlikely. The T&SD railway had been built between 1853 and 1859, and that railway runs along the very edge of Hyren quarry. I can't imagine that the railway's builders would have been happy to have people dynamiting the rock directly underneath their track. Also, from other sources we know for certain that on 14 October 1861 both James Sellick and his brother Edwin were working at Rumple Quarry when there was a fatal accident. James was injured and the inquest was partly held at 'the Riverford Inn' according to the official report.

Although the 1861 census has a different name for Riverford House, it at least is clear about the cottage, naming it 'Cottage by' Ford Farm, and this was inhabited by the Moses family, Sellick's daughter and son-in-law mentioned previously. The census of 1871 is much less helpful, simply listing two dwellings under the name Riverford. Francis Sellick, his wife Harriott, sons James, 35, and Edwin, 25, are still living at one of these, which must be Riverford House, and he still describes himself as a farmer of 15 acres. Therefore, logically the other inhabitants of the area must have been living at Riverford Cottage and they were Edward Gullett, 31, a tin miner, his wife Susanna, 27, and their family. There is no further mention of Hyren Quarry and Sellick's two sons now describe themselves as 'Farmer's sons', so we must assume it was by now definitely disused.

At Common Wood we find one person of interest among all the farmers and farm workers, William Gullett, a 61-year-old tin miner and his family. His son Henry was also a tin miner, aged 22. William was the father of the above-mentioned Edward Gullett; all the family had previously lived at Heathdown Cottage (now the site of Woodside Animal Rescue) and worked at Wheal Sidney tin mine nearby. The Gullett family feature heavily in various parts of the local history.

Between 1871 and 1881 there were substantial changes at Riverford; Vallack died in 1877 and the property was sold. The 1885 Tithe revaluation shows that most of the area, owned previously by Vallack, was now owned by Charles Norrington. However, on an adjoining rectangle of land previously owned by Erving Clarke a terrace of four small houses had been built, behind Riverford House and beside Riverford Cottage, alongside one of the access tracks. They had been built by one William Frederick Corber, but they belonged to Richard Northmore. Whereas Riverford House and the detached Cottage were quite large properties, people who knew them when they lived in the area in the twentieth century described the cottages in this terrace as very small. The ones at each end were two up-two down; the middle ones were one up-one down. Having studied the remains of these cottages, the terrace of four is not much bigger than the single, detached cottage.

In 1881 two more cottages had by now been built above, near the P&SDR. These were Elizabeth Cottage and Rowes Cottage; they were all owned by Benjamin Bromley. The inhabitant of Elizabeth Cottage in 1885 was one George Geake, of whom more anon.

By now there were a total of six families living at Riverford, but the census is again less than helpful, calling all the properties 'Riverford'. William Chapple, 40, is described as a farmer of 84 acres, so we must assume he was living in Riverford House with his wife Louisa, 36, and their family.

Edward Gullett was still there with his family which then totalled eight, so we must assume he was still living in the large detached Cottage. He was still a tin miner.

That leaves us with four families who must have been living in the terrace. Of these, two were involved in quarrying or mining; William Henry Gullett, 24, a stone quarryman and his grandfather William Gullett, 70 and still working as a tin miner. This William, as previously mentioned, was father to Edward.

Whatever the order, the families, totalling twenty people, were living in these four tiny cottages and life must have been very cramped for them.

Riverford

There was no electricity and water came from a well further up the track. As for sanitation, it's probably best not to enquire too closely! Being on a north-facing slope in a deep valley, sunshine was a rarity and the area would have been damp and misty. At least there would have been plenty of wood nearby for fuel.

The 1891 census is more specific and tells us who lives where.

At Riverford House we now have Edward Gullett, 49, and his family, and for the first time he is described as a quarryman. We know from other sources that he had been operating Colwill Quarry since at least 1885. John Lock, 27, and Francis Johns, 23, both quarrymen and presumably working for him, live with them.

At the Cottage lived Thomas Netherton, 48, an agricultural labourer, his wife Jane and their family.

Three of the four terraced cottages are occupied, although none of the occupants was involved in local industry, with the possible exception of two who worked on the railway.

The 1901 census gives us no clues about who lives where, simply listing seven families. Using information gleaned from interviews of people who had lived there in the early twentieth century, we do know who lived at the house and detached cottage; the rest is guesswork.

At Riverford House were Henry Ebdon, 50, a railway ganger, his wife Eliza, 43, and Albert Miller, 39, described as a visitor.

At the cottage were George Geake and his wife Elizabeth, both 38. George was a railway packer.

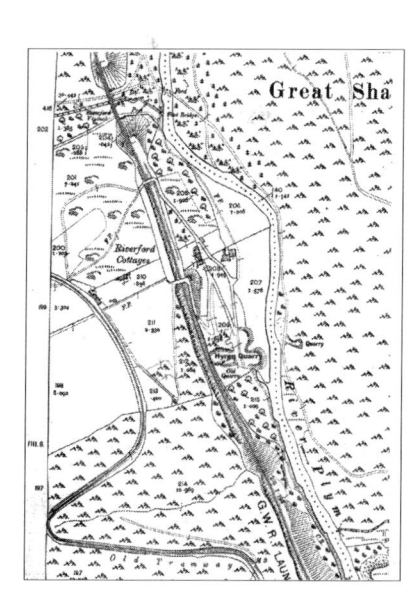

A 1907 map of Riverford. The area marked 'Riverford Cottages' is actually the site of Elizabeth Cottage.

Living in the terrace were nineteen people, only six of whom were children, so once more life was very cramped there. Of these, only William H. Gullett, 31, who called himself a roadstone quarryman and was the son of Edward, and Albert Dymond, 27, a stone quarryman, were involved in the local industry.

Henry Ebdon's story is interesting; he was born in Colaton Raleigh, south of Exeter, in 1851. By the age of 10 he was already working as a farm labourer, so obviously the family were not wealthy. By 1881 he was working as a platelayer on the railways. In 1888 he was living at Cann when he married Eliza P. Pengelley at Plympton St Mary. A deed of sale dated 2 September 1901 records the transfer of the ownership of the four terraced cottages from William Taylor to Henry Ebdon, so he had done very well for himself. Taylor had inherited them from his mother, Grace Taylor, who had herself inherited them from Richard Northmore.

The House and Cottage in 1901 were owned by Alfred Norrington, although in 1919 a further deed of sale records the transfer of these and all the surrounding land to Ebdon, so he then owned all the buildings and the land up to the GWR railway line, plus the meadow above between that line and the by then disused P&DR. The adjoining lands were owned by a variety of people but to the south the land was still part of the Colwill Estate, then owned by the wonderfully-named Pollexfen Colmore Coplestone Radcliffe, who lived at Derriford House, more or less where the hospital is now.

Eliza, Ebdon's wife, had been born in Plymouth, but her parents seem to have had a peripatetic existence. Her father Samuel was born in Ilsington, near Newton Abbot; her mother Harriett at Drewsteignton; her brothers were born at Teignmouth. In 1881 she was living with her parents in Teignmouth; also there was their grandson William Pengelley, aged 8, who was the son of Eliza.

William Pengelley was the father of Bert and Margaret Pengelley, and the family had lived first in Riverford House with his mother, Eliza Ebdon. Later they lived in the end cottage of the terrace, but William died in 1920. Margaret, when interviewed in the 1980s, said that they and their mother were threatened with eviction by her grandmother, Mrs Ebdon, but she couldn't get them out until Bert was 14 in 1922. At about this time the four tiny cottages were converted into two.

Henry Ebdon died in 1924. All the properties were then sold to Pollexfen Copplestone Colmore Radcliffe and became part of his Colwill Estate. The 1938 Land Tax return shows his widow (he died in 1930) as owning the whole area. The Walters family were at that time living at the House; Enid Walters (later Mrs Doddridge) says her father, Richard Walters, took over the tenancy when Mrs Ebdon moved out.

Enid Walters married Stanley Doddridge in 1933. Soon afterwards her father Richard Walters died; a tree fell on his leg, which had to be amputated, and he died from septicaemia. She and her husband took over running the farm. George Geake had died in 1918 but his widow Elizabeth was still in the large Cottage and she sold teas in her garden there until she died in 1937. Mrs Fry, Enid Walters' sister, recalled, 'The cottage was whitewashed and there were stone slabs, and further over were the tables where people sat. The garden was fenced in and was immaculate.'

When Mrs Geake died, the Doddridges started selling teas at Riverford House using butter churned from the milk of their own cattle and homemade jams. They also let people camp in the meadow during the summer. Mr Doddridge died of a heart attack at Newton Abbott in 1938, leaving Mrs Doddridge to carry on with help from her mother and sister, Mrs Joyce Fry. People I have spoken to who knew the area well in those days all spoke kindly of Mrs Doddridge. William Ellis, who used to camp there as a boy, recalled her taking them in and caring for them when their tent was swamped in heavy rain.

The two cottages which now made up the terrace were inhabited by various people, mainly as weekend cottages, until the early war years. John Luscombe's grandparents lived in the right-hand one to escape the bombing of Plymouth; they stayed there until about 1942, but moved out because of its disrepair. His uncle Ken rented the left-hand one, which he used as a weekend cottage, and stayed there for a while longer. Eventually the cottages became so tumbledown that they were abandoned and allowed to decay and by 1945 they were described as derelict. The Murdoch family lived in the detached cottage from 1945 until 1952. Soon after they left, possibly in 1954, Mrs Joyce Fry and Mrs Enid Doddridge, who were the final inhabitants of Riverford House, also left.

The Colwill Estate was sold by Mrs Radcliffe in 1946 and Riverford was the ninth of nine lots included in that sale. It was bought by Isaac Foot and, according to Mrs Fry, he wasn't too interested in the rent. She remembered, 'Enid used to pay the rent and I don't think she paid for years, he never used to mention it when he came out often for tea ... I think they must have been a moneyed family ... but they were more interested in the place and the surroundings, he didn't give two hoots about the rent.' She said he did little towards the upkeep of the properties and seemed quite content for nature to reclaim the land. After his death in 1960, the estate was put up for sale again and it was eventually bought by the National Trust in 1968.

So what remains today of all this activity? A casual observer would say there is absolutely nothing there, but in the bushes it is possible to find the outlines of the various buildings. Beside the first left turn when ascending the zig-zag path is the remains of the large cottage, which faced downriver towards Cann Viaduct. It was about 15 feet wide by 30 feet long and there was a shed of some kind against the retaining wall beside it. Further up the other path, now much overgrown, which

Industrial Archaeology of the Plym Valley

Above left: Riverford today.

Above right: The iron bridge over the railway track.

ran from the cottage up to the iron bridge are the remains of the terrace. This was about 20 feet wide by 36 feet long but the ends are too tumbledown to make out. The fact that four dwellings take up little more space than the single one shows just how small they were. There could be buildings at the ends, a wall around the front to possibly provide a garden and a hole about five yards further along the track which I assume was the well.

Of Riverford House, the largest of the buildings, very little remains. The garden wall is still there, plus a few remnants of odd parts of the house walls, but considering how large it was there is very little to show. Presumably the stones were all removed for use elsewhere, because even the stones from a house that size would form a large pile if they were still in the area.

The most obvious signs of all this human activity are the access tracks, both of which are still there, as are the bridges which allowed the tracks to cross the railway line above, and of course the quarry. Surprisingly, the remains of the inclined plane which served Hyren Quarry still exist between the railway line (now a cyclepath) and the tramway above. With map in hand, I have seen where a depression crosses the top end of the zig-zag path and runs down to the railway cutting, exactly where the inclined plane was.

Whereas Riverford has disappeared, the cottages at the top of the hill have prospered. There are now four of them, all privately owned by people who are jealous of their privacy. Elizabeth Cottage stands beside the remains of the P&D railway track. Rowes Cottage below it has become a large house, now called Odoorn Lodge. Common Wood House is next to Elizabeth along the track and further along again is Common Wood Cottage. Their access road is the remains of the original Stuckey Lane.

PLYMBRIDGE SLATE QUARRIES

The valley of the River Plym is an area of natural beauty on the outskirts of Plymouth, and as such is immensely popular as a place to walk and relax. The valley is steep-sided, deeply forested and the ground has that furry, mossy look of an ancient woodland – a perfect place to explore.

Plymbridge is the focal point for much of this activity, and a sunny weekend in summer will see dozens of cars parked around the area, in various car parks and all along the roads in all directions. Three-quarters of a mile north of the National Trust car park at Plymbridge, visitors, whether they are walkers or cyclists, come upon a different sight, a large viaduct striding majestically across the wooded valley. This viaduct provides a perfect vantage point, from which can be seen two large pits, one each side of the river, with occasional ruined structures also visible, the whole area much overgrown now as nature reclaims its own. The pit on the right (eastern) bank is best known these days as home to peregrine falcons, kestrels and ravens; obviously these pits were previously evidence of man's activity. What were they?

This part of the Plym Valley, between Plymbridge and Riverford viaduct, was an area where large outcrops of slate and blue elvan stone were found. Despite some of the claims at the time the slate apparently was not of the very highest quality, but was certainly good enough for roofing and flooring; the elvan was a hard stone which is very useful for road building.

Cann Quarry

Quarrying for these useful materials was probably carried out from very early times; there are many small-scale pits dug into hillsides hereabouts. Certainly records exist from Plympton Free School Repairs Account Book, dated 10 October 1671, recording payments to John Hill and John Lyde for '"healing stones" from Cann quarry'. (Helling or healing stones are a local term for flat stones for roofing).

Cann is the pit on the eastern side of the river, on land which at that time was owned by the Parker family (Barons Boringdon from 1784, becoming Earls of Morley from 1815), and was one of the larger quarries. The first known lease, between George Parker of Boringdon and Nicholas Edwards the Younger of Plympton, is dated 29 September 1683. This lease was to run for 7 years with a rent of £6 annually, to be paid quarterly. It covered 'All that helling stone quarry known as Cann Wood Quarry', and George Parker made sure that his needs would be covered – he was to be supplied with 'slatt or helling stone' for repairs to Boringdon Hall, stables, barns and outhouses at rate of 4s per thousand stones. Also Paviers (paving stones) as required. Edwards even had to find the workmen – hellyers for slating, and also men for pointing, tiling, low-casting and plastering. Parker was to provide these workmen with food and drink and pay 10d per man per day's work.

During the eighteenth century Cann Quarry was operating under various leaseholders or else under direct control of the Morley estate. A lease of March 1781 mentions 'Cann Quarry, in the Parish of Plympton St Mary, in the County of Devon, with the dwelling house and buildings thereto belonging and adjoining.' This is the first mention of buildings attached to the quarry – the dwelling house refers to what was known as the 'Foreman's House', later Cann Cottage, which is now more generally known as Railway Cottage.

In 1795 there was a new lease between John, Lord Boringdon, and Richard Drew, a printer of Kingsbridge, which includes the instruction, 'A new head of 9,000 feet to be broken.' We must presume that the printer Drew knew little about quarrying because the Morley accounts show him owing £85 between 1809 and 1815, during which time Thomas Gullett is leasing the place and he also has the instruction to 'break a new head on the southern side' within three years. The family name Gullett is one which appears frequently in the history of quarrying and mining in the valley.

In 1797 the Revd John Swete, a noted traveller of the time, visited the area and sketched Cann, although it is difficult to reconcile this sketch with how the quarry looks today. A later, undated, engraving shows Cann Quarry looking much more like a working quarry and gives us an idea of how it looked before the arrival of the railway and the building of Cann House.

Plymbridge Slate Quarries

A sketch of Cann by the Revd Swete.

Engraving of Cann.

By 1816 there is no record on the estate accounts so Thomas Gullett may have given up, and by 1818 the lease is with Knapman & Co., but according to an agreement dated 1821, 'Cann Quarry aforesaid and the Quarry House and Smith's Shop thereto belonging, now respectively in the tenure of William Knapman and James Gullett or one of them', so Thomas' son James was certainly still involved with the quarry. In 1824 there was apparently some dispute between Knapman and the, by then, Lord Morley because the lease is abruptly terminated, the matter goes to arbitration and Lord Morley has to pay Knapman £1,376 4s 3d, plus lawyers' fees totalling £84 13s 10d and witnesses charges of £16 3s 2d. This was an enormous sum for those days, equivalent to roughly £1.2 million today when compared to average earnings.

Even today the area is served only by minor roads, so obviously transporting the stone would always have presented a problem. Originally, the only way to transport the stone from the quarries in this area was by rough tracks to Plymbridge and thence along the narrow lanes to Crabtree, at that time the nearest tidal wharf, or else onwards by road. In 1823, when Sir Thomas Tyrwhitt's Plymouth & Dartmoor Railway Company (P&DR) opened its line down through the opposite side of the valley, the removal of slate from at least some of the local quarries was made much easier, but this was of no benefit to Cann.

The leat which supplied water to the Marsh Mill had run from Cann Quarry parallel to the river since at least 1723. Smeaton (who built the lighthouse which used to be on the Eddystone Reef and is now on Plymouth Hoe) surveyed the area with the intention to convert the leat into a canal for transporting the slate as early as 1778. However, when the P&DR was being built they needed to pass over some of Lord Boringdon's land and part of the deal agreed was that the P&DR would, at their expense, connect Cann Quarry to their tramway. As the tramway at its nearest is 100 feet above the quarry and on the other side of the valley, this was a ridiculous deal to have agreed. It would have been at the very least an enormously tricky and expensive operation. When faced with these problems the company initially refused to comply with the agreement, but when two of the three directors who had agreed to the deal, Sir William Elford and Mr John Tingcombe, were bankrupted by the collapse of the Plymouth Bank, leaving Mr John Pridham faced with having to pay Lord Morley a penalty of £5,000, the company caved in and agreed to fund whatever was needed.

Rather than connect directly up the hill to the P&DR, it was agreed to connect with the tramway at Marsh Mills. John Meadows Rendel surveyed the land and suggested a tramway to run parallel to the river from Cann to Marsh Mill but John Parker, who by now had been created Earl of Morley, still preferred a canal and so the leat was widened and strengthened between 1827 and 1829, when the canal started carrying stone. The canal was short-lived; having to load slate into barges at Cann, then unload it again and transfer it to wagons at Marsh Mills for onward shipment was far more trouble than it was worth. Certainly by 1839 and probably much earlier, the P&DR was extended to Cann along the towpath of the canal as Rendell had originally suggested. The canal reverted to being simply the mill leat again, if an enormously wide one. All of this work was done at the expense of the P&DR!

As an aside, in the process of constructing the canal, a lode of lead and silver-bearing rock was discovered in the hill beside the canal. The Cann Mine, otherwise called Canal Mine, was set up to extract this ore between 1824 and 1825.

A lease reversion dated 1828 mentions 'that newly erected dwelling house and outhouses, garden and orchard'. This is the first mention of what was

Plymbridge Slate Quarries

variously referred to as The Cottage, Cann House or Cann Quarry House, built as the suitably grand dwelling of the owner or manager. The walls of this building are still visible in the woods to the south of the quarry and it is depicted in an etching dated 1831, showing it to have been what an estate agent would now call a large dormer bungalow in idyllic surroundings, with gardens around it and outhouses at the rear. On flat ground below Cann House are shown the 'Foreman's House', now known as Railway Cottages, with other buildings dotted around, one of which is the Blacksmith's shop. In front of these is a meadow for grazing animals and other buildings, probably stables and barns. As this pre-dates the arrival of the Tavistock & South Devon (later Great Western) Railway, all these buildings lie on artificially flattened areas carved out of what was otherwise a hillside. The quarry itself is not directly visible in this etching.

An etching of Cann House.

The history of Cann is littered with leases taken out and shortly afterwards being offered again because the grandiose ideas of the leaseholders did not match the reality. Having read many of the adverts offering the place for lease, I can't help feeling that had there been a Trade Descriptions Act in those days there might have been prosecutions! Here is one example.

11th December 1834   TREWMAN'S EXETER FLYING POST

VALUABLE SLATE PROPERTY
DEVONSHIRE

To be sold by Private Contract, the CANN SLATE QUARRIES, Plympton, Devon, held under Lord Morley for a term of which about 90 years are unexpired.

The property contains an inexhaustible quantity of the most valuable slate of the very first quality for durability and beauty. [Author's emphasis]

The Quarries are within 4 miles of Plymouth, by which they communicate partly by a private canal of the proprietors, and partly by rail roads, terminating at a Wharf and Quay at Plymouth, and a Quay at Cattwater, in deep water, capable of admitting ships of any burthen, and where vessels load free of all dues.

There is a picturesque and roomy Dwelling House, Lawn, Orchards, Garden, and Stabling, on the property, and several Workmen's Houses.

Upwards of 30,000 pounds has been expended on the works during the last few years, and, to an individual or company, the property affords advantages seldom to be met with. The purchase money may, if required, be paid by instalments at extended periods.

Who could afford to miss out on such an opportunity? After three years trying, and failing, to sell or auction it, in 1839 it was once more leased. This lease for Cann was agreed between John, Earl of Morley, and George Reeves. This is the first lease which includes a map, albeit a sketched one, showing the layout of the quarry.

Plymbridge Slate Quarries

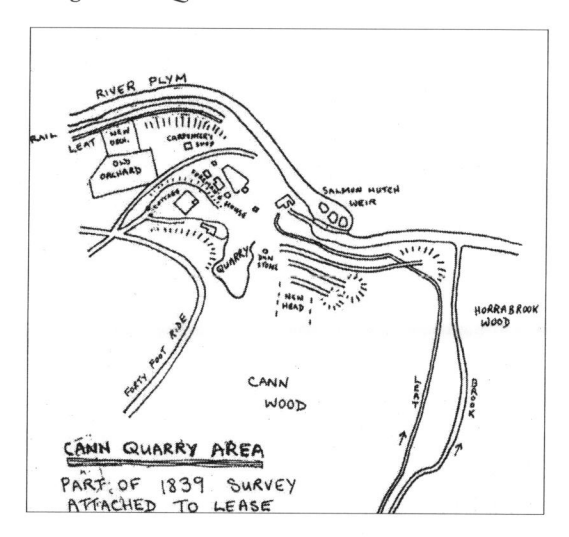

1839 plan of Cann.

New Head.

The area to the north of the major part of the quarry is marked 'New Head', so somebody was finally about to open the new head, even if it was on the opposite side to the original site. The New Head was presumably a venture looking for more slate, but it appears not to have been a success. It is a long, narrow quarry (roughly 160 feet long and 75 feet wide according to Google Earth) beside the existing quarry, which is roughly 355 feet long and 260 wide. A huge spoil tip from the main quarry has partly filled it, so it has been disused for a very long time.

The map also shows the mill leat/canal plus the railway beside it, and also the separate leat for the overshot waterwheel running down from a reservoir on the Horrabrook (the stream which joins the river about a quarter of a mile north of the weir), the foreman's house (which we now know as Railway Cottages), the carpenter's shop and the 'cottage', which we now know as Cann House,

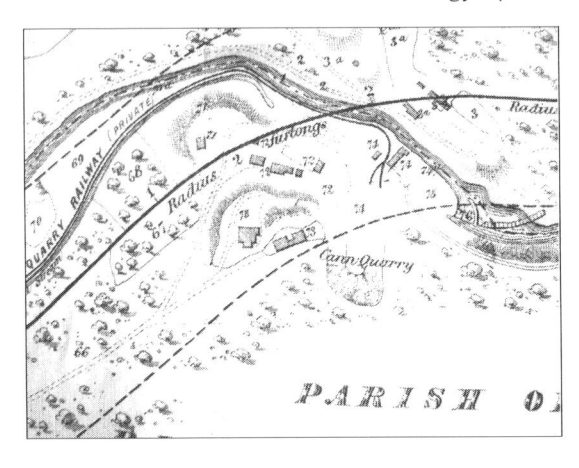

Cann, as shown on the 1853 T&SD plan.

plus the orchards mentioned in the lease. Down in the quarry itself, there are other buildings apart from the waterwheel – workshops where cleaning and preparing of the slate could be carried out, including cutting and planing the stone using machinery powered by the waterwheel. Tracks lead north from the quarry to spoil tips beside the river.

In 1853 the Tavistock & South Devon Railway published a plan showing the intended route of their track, and this shows Cann at that time. This is a real map as opposed to a sketch.

The 1841 census shows Eusabia Martin, 35 and a quarryman, of whom more anon, his wife Mary, 25, and their daughter Susannah, 4 months, living at 'Cann Quarry', which would have meant the Foreman's House/Railway Cottages.

A lease of 1845 between Lord Morley and one Charles Pike mentions 'buildings, sawing and planing sheds, wooden water wheel'. Mention is made of 'machinery, engines, capstan, ropes, tackle, rails, wagons, trams, tools, etc'. This is also the first time that elvan stone is mentioned in a list of tonnage rates; 6 pence for slate and 1 penny for elvan stone, refuse and rubble.

The year 1849 saw a new lease for Cann and the arrival of one Thomas Pearson, who must have cut quite a dash when he arrived. He appears to have been the first person to actually live in Cann House, and the 1851 census shows him, his wife Louisa, 36, and their daughters Clara Louisa, 5, and Mary Jane, 3, installed there. Pearson grandly described himself as a 'Merchant proprietor of Slate Quarries, Houses, etc.' His 17-year-old brother Henry, described as a clerk, lived with them as did Rebecca Bond, 48, a cook, Mary Ann Kelley, 19, a housemaid, and Henry Stanton Ruse, 21, their coachman. Unfortunately, his income appears not to have kept up with his grandiose lifestyle because by July

Plymbridge Slate Quarries

1851 he was proclaimed bankrupt with the then enormous debt of £6,800! The lease of the quarry reverted to Charles Pike, who worked on until the late 1850s.

Apart from the Pearson family and entourage, the 1851 census for Cann also shows Abraham Bellamy, a blacksmith, his wife and four children, plus John Martin, quarry labourer, his wife, her brother and their grandson, plus two more labourers as lodgers, living at what had previously been called the foreman's house. It must have been very cosy! Incidentally, John Martin and his wife had been living at Riverford, and presumably working at Hyren quarry, ten years previously.

In contrast, by 1861 there were only two occupants in the old Foreman's house. These were Thomas Harris, a Chelsea pensioner, and his wife, so nobody connected with the quarry was living there at that time. This was presumably one of Cann's lean periods.

In 1863 Cann and Rumple quarries were taken over by the Plym River Slab & Slate Co., which is covered more fully under the Rumple heading.

After the demise of the above-named company in 1868, Cann would again have been quiet. The census for 1871 shows 55-year-old Thomas Harris, an agricultural labourer, his wife and family living at 'Cann Quarry House'. This seems unlikely to be Cann House – it is more likely to be Railway Cottage but we have no proof.

In 1874 a lease for Cann is proposed, giving the operation of the quarry jointly to James Soper, who 20 years previously had run Rumple, and James Gullett. The lease was to run for seven years and when it expired in 1881 James Gullett, by then 78, wished to retire and surrender his rights to Soper. Soper carried on until 1885, when he sold up all his plant.

The 1881 census shows the Cann area much more densely populated than had been the case ten years before. Cann Quarry, presumably meaning Railway Cottage, had two families living there: William Gully, his wife, six children and a lodger, plus James Arthur Beavis, his wife and young daughter. All three adult males were described as Railway Packers (with the written addition 'navvy'), so they appear to have had nothing to do with the quarry.

Also listed at Cann, but presumably living in Cann House, was James Soper. He now describes himself as 'Master of slate quarry employing 4 men and 1 boy'. With him were his wife and two children, and listed separately as head was Robert Sorton, a clay labourer, and his three children.

In 1891 some members of the Royal Naval Engineering College photographic club visited the area and took pictures, including one taken from across the river showing the waterwheel and buildings.

Industrial Archaeology of the Plym Valley

Above left: 1891 RNEC picture of Cann waterwheel.

Above right: Enlargement of of 1891 picture showing buildings.

Enlarging this picture, it is possible to see that already the buildings are looking shabby and in need of repair. On the right of the picture above the roof of the left-hand building, where a stone plinth still exists, is a structure which could be some form of crane.

The 1891 census still shows James Soper living at Cann House, although he appears not to have been involved in running it. He was by then 71 years of age and describes himself as a Slate Merchant. He was apparently the final occupant of Cann House. The occupiers of Railway Cottages are all railway workers.

Cann's last operator was Edward Gullett, who as far as I can tell was a nephew of his predecessor James Gullett. He was living at Railway Cottage in 1901 with his wife Susannah and disabled son Harry, daughter Edith and Edward's sister Mary (known as Polly). Also with him were his twin grandsons, Francis and Percy Lock, and two lodgers. Edward and the two lodgers described themselves as 'Roadstone Quarrymen' because by now the slate was all worked out and they were producing only blue elvan stone. Albert Gullett, another son of Edward, and Jack Maker also worked in the quarry.

Why was Edward living in the smaller Railway Cottage rather than the spacious Cann House? Could it, I wonder, have anything to do with the rent? Back in 1874 Lord Morley had asked £5 per annum, which Soper seems to have been happy to pay, but no doubt it would have been much higher 27 years later. Edward Gullett strikes me as being a no-nonsense, hard working, canny man, a man of substance rather than show who, given the choice, was happy to pay only what he had to and live in a smaller, if less comfortable, house.

Plymbridge Slate Quarries

Above left: Edward Gullett in his smithy.

Above centre: Polly Gullett.

Above right: Harry Gullett.

Edward and Susannah worked hard; as well as running the quarry, he would keep pigs and slaughter them for the bacon and she would buy fish, wash them in the canal and salt them down for winter, so the quarry obviously wasn't making them enormously wealthy. They also, like many in the valley, provided teas for the many day-trippers who walked in the area at weekends, and a picture shows the tables beside the cottage for the customers. Edward retired in about 1910, but until then he carried on producing elvan stone, which he sent down the Cann Quarry branch line in trucks drawn by Clydesdale horses to the council stone-crushing plant near the Marsh Mill for use on the roads.

After his retirement he, Susannah, Harry and Aunt Polly still lived at Railway Cottage until his death in 1915. The rest of the family stayed there until 1923. Harry built a waterwheel, using the old waterwheel house and machinery, to generate electricity for them. The remains of the waterwheel axle and gearing are still visible today. Albert Gullett emigrated to Australia in 1920.

Since then, the cottage was inhabited by various people, becoming increasingly tumbledown up to about 1955, when it was left to decay. Isaac Foot, who by this time had become the landlord, died in 1960 and the cottage, as part of his estate, was sold to the National Trust in 1968. Cann House, having been disused for much longer, is now little more than a ruin, a sad end to what must have been a lovely house.

Above: Railway Cottage auction picture.

Left: Cann waterwheel remains.

Rumple Quarry

Rumple opened slightly later than Cann; the following advert appeared on 8 June 1761 in the *Sherborne & Yeovil Mercury*.

> This is to give notice that Rumple Quarry, in the Parish of Egg Buckland, in the County of Devon, late in the possession of Robert Warn, and now of William Stevens, of Crabtree, is just opened.

Warn and Stevens were lessees, the landowner was Henry Tolcher, Lord of the Manor of Eggbuckland. He died in 1818, whereupon his nephew Christopher inherited the estate.

In 1838 Rumple Quarry was taken over by Messrs Kingwell, Conway and Symons, who started advertising their wares locally. By 1841 Robert Gullett and George Moon, described as slate quarry labourers, are living with their families in the cottage which stood to the right of the quarry entrance.

Things changed greatly over the next twenty years. The 1850 Poor Law Rate Book for Eggbuckland shows James Soper as operator and Robert Gullett still living in the cottage along with Eusabia Martin, another quarryman who had previously worked across the river at Cann. However, by the time of the next census in 1851 James Soper, 31, his wife Agnes (née Gullett) and their 5-year-old son Robert were living in the cottage and Thomas Cowling, a 44-year-old quarry worker, was lodging with them. The census gives Soper's occupation as 'head of slate quarry employing 13 men'. To complicate things further, the

Plymbridge Slate Quarries

1853 deposited plan for the Tavistock & South Devon railway shows Abraham Bellamy, a blacksmith who had also previously worked at Cann, living at the cottage and the lessees of the quarry as James and George Soper. These two were sons of James Soper the elder, also a stonemason, and my feeling is that he, rather than his sons, would have been the original named operator.

Between 1853 and 1859 the T&SD railway was constructed along the valley, including the viaduct which crossed the river here. On the Rumple side the supports for the wooden viaduct straddled the existing workshops and stood close beside the cottage, but the 1861 census shows James Soper and his growing family, now two sons and two daughters, still living there. Shortly after the census Soper stopped operating the quarry, which was taken on by a Mr William Weekes. The 1865 Ordnance Survey map shows the cottage still in position. There is no census record for Rumple in 1871 – the cottage was demolished sometime between 1866 (when the 25" OS map was published) and 1884 (when the 6 inch map was published).

Also shown on this map is a small footbridge straddling the river beside the viaduct. This would have been essential to save a very long walk to get

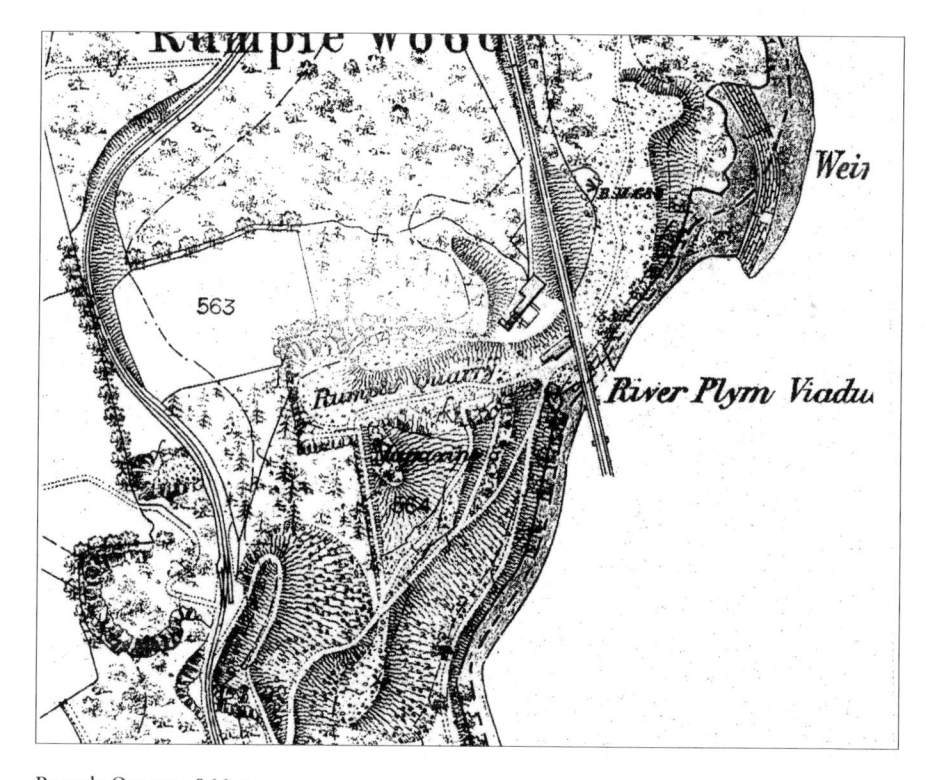

Rumple Quarry 1866 map.

Industrial Archaeology of the Plym Valley

from one quarry to the other, as at this time both were being operated in tandem.

The 1866 map also shows Rumple quarry in detail. Soper, or a predecessor, had come up with an unusual solution to the problem of disposal of waste rock, known as spoil. Because Rumple is long and narrow (roughly 330 feet long by 135 feet wide), and ends beside the riverside, getting spoil out from the quarry face had obviously been a problem. Their solution was to make a cutting through the Southern wall of the quarry, about two thirds of the way in, and take the spoil out there. This is shown on the 1866 map, as is another track which appeared to come up from the quarry face along the side of the quarry to the cutting. From the end of this cutting, the spoil was dumped down the slope of the existing hillside. This enormous pile of slate spoil, very visible on the left as a visitor walks along the western bank of the Plym about 120 yards before the viaduct, is one of the more obvious signs of the amount of work carried out in the area. Although the cutting is still in place, the path along the side of the quarry had been lost by the time of the 1904 map.

October 1861 saw a 'Frightful Accident' at Rumple. Eusabia Martin and Maximilian Winsborough, along with several other men, were working in the quarry. There had been some blasting done 2 hours previously but it had not dislodged some stone at the very top of the quarry. Later, while the men were working underneath, this large area of stone fell. Martin and Winsborough

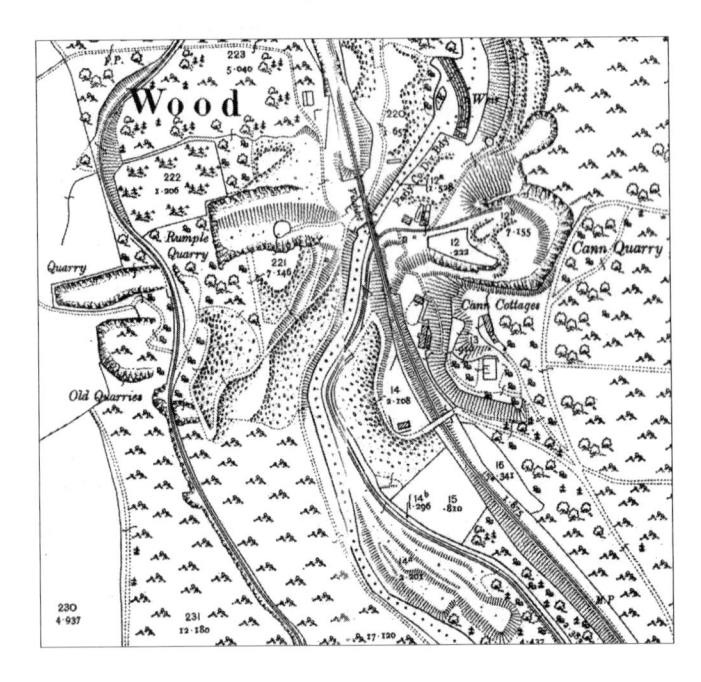

1904 map of Cann and Rumple.

were killed and three others were injured, one of whom was James Sellick, the blacksmith son of the then inhabitant of Riverford House. His brother Edwin, 17, was an eyewitness. At the inquest James Soper, who had given up his tenancy of the quarry seven months previously after working it for eleven years, said that Martin was a very experienced quarryman and the work was carried out on his (Martin's) orders. Soper said he had not been happy with the work when he last saw it because there was an overhang at the top. It was this overhang which had fallen. The inquest verdict was Accidental Death.

Soon after this, Rumple and Cann were both taken over by the Plym River Slab & Slate Company (PRS&S). This was an impressive-sounding company based in London which issued a prospectus which was reported in the *Western Daily Mercury* of 27 November 1863 as follows:

> This company is formed for the purchasing and working on an extensive scale, the old and well known Rumple and Cann slate quarries, situate on the River Plym, about 5 miles from the naval town of Plymouth. The property comprises about 200 acres of *slate rock of an immense thickness, which improves in quality as it increases in depth*; and the leases, plant, buildings, etc., and all existing rights therein, have been purchased on favourable terms. *The slate is practically inexhaustible in quantity, and of a superior quality*; is of good colour, solid, strong, and durable, and resists well the influences of the atmosphere and fire. It can be quarried in blocks of large size, and is admirably suited for roofing, cisterns, chimney pieces and the innumerable other purposes for which slate is generally used. The colour quality and durability is equal to the yield of the far-famed Delabole and Welsh quarries.
>
> The property has been inspected by several gentlemen eminent for their practical knowledge of slate quarries, and their reports furnish full particulars of the facts and data on which their estimates of the probable profits of the company are based. *From these it will be seen that an early return of 30 to 40 per cent of profits will be realized, with a prospective increase as the works are opened up.* [Author's emphasis]

Now maybe I'm an old cynic, but to me this simply screams 'Con Artists!' The shares were £6 each, 10,000 were issued and they seem to have been well received, so maybe they were more gullible in those days.

At about this time ownership of the Colwill Estate, of which Rumple was a part, passed from Christopher Tolcher, first to Mr John Bennett and then to Coplestone Lopes Radcliffe, who also owned the nearby Derriford Estate. Tolcher has been described as 'rather dissolute and spendthrift', which may explain the sale. Bennett owned the estate in 1864, because an Indenture dated 10 May of that year grants the rights to 'dig, quarry and search for' slate to the Plym River Slab & Slate Co. Ltd, of which he was also Chief Engineer and a shareholder. It appears Mr Bennett took out a mortgage with Mr Radcliffe in order to buy the estate then another mortgage with him in August 1866, bringing the total money owed to £4,000, presumably in an effort to stave off bankruptcy. In this he failed, because in September 1866 he was made bankrupt and the estate passed to Mr Radcliffe.

One result of the operation of both Cann and Rumple quarries by the PRS&S was the building of a waterwheel to power Rumple. This was opened with great fanfare in September 1865 and is described in a long article in the *Western Daily Mercury*. There had been much speculation as to whether it could be made to work because it used water from the same reservoir on the Horrabrook as the smaller, and hence much lower, waterwheel across the river in Cann quarry. The engineer, Mr John Bennett, who has been mentioned above, was confident however and at the grand opening the water came, albeit slowly and described as 'a little stream', but the wheel did turn at a rate of eight revolutions per minute. There was much eating, drinking and jollity to celebrate this, but it is doubtful whether the wheel could have produced much power.

A strike by the Amalgamated Society of Engineers during 1864 and 1865 affected all branches of construction. There is no documentary evidence to say that the PRS&S Co. was affected but it must have been so. Whether this had any bearing on the company's fortunes is therefore open to speculation, but by July 1865 shares were being offered at a substantial discount, up to 50 per cent. As early as January 1866, the following letter from a disgruntled shareholder appeared in the *Western Daily Mercury*, comparing the operation with that of another local quarry.

> Sir, I see in your valuable paper a good report of the Alexandria Quarry. I have seen the quarry and believe it to be a good one. If well and economically managed, good results will follow – not forgetting that too many cooks spoil the broth. For example, the Plymouth River and Slab Quarry [sic]. This place was worked for some time and there was a large sum of money muddled away. There are three steam engines at the quarry – two daily working, incurring the expense of engineer, coals, grease, etcetera. While they have a splendid and powerful water wheel remaining idle, with a full

supply of water, and power enough to pump and haul, making a very large saving over the present power. Surely the shareholders ought not to look on and see their money squandered. No doubt if this quarry was managed by a practical man it would pay. Instead of that it is generally managed by men who know more about army and navy than they do about quarries and unfortunate shareholders.

I am Sir
One of the latter

By mid-1866 the PRS&S was selling off its equipment and people were taking court action to recover money from them. Bankruptcy proceedings began in August 1866 and the final winding up order was issued in March 1868.
 To mis-quote Oscar Wilde's Lady Bracknell, to have one scoundrel running Cann and going bankrupt is a misfortune, two seems like carelessness.
 Recovering slate from Rumple quarry was hard work in itself; how did they get their slate to market? After 1823, when the P&DR opened above the quarry, that would have presented the easiest way of carrying the slate, but of course that meant hauling it up a steep, zig-zag path visible on the map above by the side of the quarry. There is also the track, still existing today, which runs along by the side of the river to Plymbridge. This is more or less level and might have been an easier exit route. There is no hard evidence that there was ever a tramway along this track, but recently granite setts have appeared as the track erodes. Was this their transport route, at least for a short time between map surveys?
 A far better way of getting the slate to market arrived with the building of the T&SDR, which crossed the river beside the quarry. Near the northern end of Cann Viaduct there is a building at the top of the slope which runs from Rumple up to the railway line; this was probably designed for a steam-powered haulage system to bring stone from the quarry up the slope to the railway siding. An imaginative information board nearby shows how the system would have worked. An excellent idea which maybe was built as a last gasp just before the PRS&S went bankrupt, but did it work? The 1866 25-inch OS map shows nothing there at the time and the old Rumple Cottage was still in position, partly blocking the line of the slope. The 6-inch OS map of 1886 shows the building as walls but with no chimney or roof. This map was surveyed sometime between 1859 and 1886, but other clues suggest this area was surveyed towards the latter date. The GWR deposited plan which shows the new viaducts was surveyed in 1908 and this describes the building as 'Ruin'

Rumple winch house

and the walls shown are identical to the earlier ones. My feeling is that the walls were built and the rock at the back was cut away to allow the flue to run up the hill to a chimney – even the round foundation for the chimney was cut. Then the money ran out! All of this is still visible today. The most expensive part would have been the steam engine itself, and I believe that was never fitted, nor was the chimney or flue ever built; there is no sign of either having been built and subsequently demolished.

When did Rumple close? This is open to conjecture. There is evidence that it closed in 1868, after the demise of the PRS&S Co. There was a bridge spanning the river below the viaduct which was shown in the 1865 OS map, so there was the possibility of direct transport between the quarries, although the only picture I have seen of the last bridge shows a footbridge and anecdotal evidence says it was never more than that. Certainly a picture, taken of the new viaduct and hence after March 1907, shows wheel tracks in and out of Rumple and there is talk that Edward Gullett worked Rumple until the last seam of slate ran out. Thereafter, it was said to have been used as a dumping ground for spoil from Cann.

OTHER QUARRIES

There was an un-named quarry above Rumple, but that had obviously been worked out by 1823 when the P&DR opened, because an embankment is constructed right across the entrance to the quarry. The P&DR would have made transport much easier for the other quarries in this area.

Also above Rumple was Higher Rumple, of which little is known except that on 2 May 1774 the *Sherborne and Yeovil Mercury* carried an advert as follows:-

> To be let at an annual rent, from midsummer next, all that well known slate stone quarry called HIGHER RUMPLE quarry, situate in the parish of Egg Buckland, in the County of Devon, late in the possession of William Stevens, as tenant thereof, about 3 miles from Plymouth and 1½ from a navigable river; from whence the said stone may easily be carried to the said town, or Plymouth Dock. For which purpose a public survey will be held, on Wednesday the 11th day of May next, by 4 o'clock in the afternoon, at the home of Mr Richard Brooks, in Egg Buckland, aforesaid. April 26th 1774.

We must presume that this was the same William Stevens who was joint operator of Rumple in 1761.

This apart, the only other reference to this quarry is its appearance in the 1840 Eggbuckland Tithe Map. The 1865 OS map (shown above) shows a track which led directly from the quarry, around the edge of the neighbouring Colwill quarry, to Colwill farm above. This track is still in existence today, ending at the boundary with Wrigley's grounds.

We also have little information about Colwill Quarry, just a few yards along the P&DR from Higher Rumple, except that it was not on the Tithe Map, so it must not have been in operation at that time. However, in February 1890 it made the news in a big and very unfortunate way. The operator was Edward Gullett, who later operated Cann and at the time was living at Riverford. He had been operating the quarry since at least the time of the 1885 Tithe Revaluation. His nephew William worked at Colwill for him along with Edward's son, also William, George Tapper and John Lock. They used dynamite to blast the rock and one problem with dynamite is that in cold weather it becomes too hard to explode and has to be warmed until it returns to a 'pasty' state. During the warming process it can give off nitro-glycerine, a very unstable and dangerous liquid.

On the day in question, the nephew William and George Tapper were warming some sticks of dynamite ready for blasting. Their method for doing this was to put water into a large paint pot which was placed over the fire in the smithy. The dynamite sticks were put in a piece of sacking which was placed on top of the tin and the dynamite was 'steamed' until it was warm enough. I can't imagine what the Health and Safety Executive of today would have to say about this practice! The problem was that nitro-glycerine from the sticks would seep through the sacking and sink to the bottom of the water. Everything went well during the first operation but when they put a second load on to heat, the nitro-glycerine exploded; luckily it didn't set off the dynamite. Tapper was dragged from the smithy and died on the spot; William was taken to hospital and died the next day.

The inquest found the deaths to be accidental, but censured Edward Gullett because he had known what they did and condoned it. No criminal proceedings were taken because it was felt he had suffered enough with the death of his nephew and friend.

All the above quarries were in the same area and presumably working the same seam of slate, their main source of income. Later, when the slate ran out, some went on to produce elvan stone for road building. Further down the valley, however, was the small Plymbridge quarry, reputed to have been worked by Edward Gullett when he was running Cann. Just beyond Plymbridge was Mainstone Quarry, which exclusively produced elvan stone for road building. This quarry started in the late nineteenth century and continued until 1950. By then it was the only quarry still operating in the valley.

Further up the valley at Riverford, Hyren quarry was also in operation from the early 1840s, the operator being one Henry Crace. We know the stone was transported from this quarry up to the P&DR above using an inclined plane as it is mentioned in the proposal for the Tavistock & South Devon Railway (T&SD). This inclined plane was powered by a steam engine, as can be seen in the notice when it was sold in 1847.

> The steam engine is quite new and has been recently erected in the quarry and is of the best material. In addition to its being fitted for raising the slate from the quarry, it has all the requisites for being used as a saw mill, having a circular saw bench complete with machinery.

In the 1853 T&SD proposed plan Hyren is shown as being operated by Francis Sellick, who farmed the land and ran the tea room at Riverford; his eldest son George, 19, is shown in the 1851 census as a slate quarryman. This is the latest documentary evidence of the quarry being worked. Once the T&SD railway had been built, between 1856 and its opening in 1859, it ran right across the top of the quarry face. I can't imagine the railway engineers being too happy about people blowing holes in the rock face below their railway so I think it's fairly safe to assume that Hyren was abandoned by that time.

Across the river from Hyren is another quarry of similar dimensions, presumably on the same seam of slate; possibly this was simply an exploratory dig by the operators. Nothing is known of this quarry except that the 1907 OS map shows it as 'Quarry', possibly meaning it was still being operated, as opposed to Hyren which is shown as 'Old quarry'.

So what can be seen of all this today? Whether you walk up from Plymbridge along one of the riverside paths or cycle along the old railway track, now a cyclepath, the most obvious point at which to start looking for evidence of all this work is Cann Viaduct. At the northern end of the viaduct, furthest from Plymbridge, a track leads off to the left towards an old siding and a slope running down to the riverside. At the top of this track is a very strongly-constructed building which has a banjo-shaped cutting in the rock of the hillside behind it to take a flue and chimney. At the front of the building is an L-shaped wall; this is thought to have been the site of the haulage system for an inclined plane which would have been used to haul stone up the slope, ready for loading onto the railway at the siding. Beside this wall would have been the drum of a winding system, powered by the steam engine which is presumed to have been in the adjoining building. All of this is conjecture; I have seen no proof that the building was completed as described, and certainly none that it was ever used in this way, but an information board nearby shows an artist's impression of how it might have looked if it had worked.

A wall marks the position of the siding, which was alongside the original railway track; by the time the new, current viaduct was built in 1907, there would have been no reason to realign this siding as it was no longer needed. The cutting beyond this building, where the railway ran through the rock of the hillside, has obviously been widened to allow for the track realignment.

Walk down the slope and Rumple Quarry runs on your right at an angle from the riverbank into the hillside. It is possible to walk with care the whole length of the quarry, but it is muddy, with old fallen trees and boulders strewn around the floor. A recent rock-fall from the southern side of the quarry shows that these places can be dangerous.

The face of the quarry, the area which was worked, is an impressive 100-foot-high wall of rock. After heavy rains the stream which runs down from the Wrigley's factory above cascades down the face, forming a substantial and quite picturesque waterfall. No buildings remain which were associated with Rumple. The cottage in which James Soper lived stood to the left of the slope, roughly two-thirds of the way down, just before the quarry entrance. It ran at an angle towards the embankment, as can be seen from the 1866 map above. It survived the wooden viaduct but not the second, stone one. The building which stood in front of the cottage, possibly used as a store or workshop, is gone but fragments of wall remain beside the riverside path. The whole of the slope on which these two buildings stood has been transformed by the dumping of huge amounts of spoil, extending from halfway along the northern side of the quarry across to the embankment of the new viaduct and continuing around the other side as well.

Industrial Archaeology of the Plym Valley

Above left: Rumple waterwheel building from Cann side of the river now.

Above right: Rumple wheel pit from above.

Following the riverside path around below the viaduct, the remains of the substantial waterwheel building are on the right of the track, with an information board nearby. The line of the building faces across the river towards the end of the leat on the hillside opposite, now hidden behind trees. The leat supplying water for this wheel would have been carried to the wheel on a wooden aqueduct called a launder.

Retracing our steps back under the viaduct and past the entrance of the quarry, a track splits off from the riverside path and climbs up the hillside. Above you on the right at this point are the remains of a square building which was shown on the OS map as the magazine. A hundred yards up this path a level track joins from a cutting on the right; this was the cutting through which the spoil from the quarry was brought to be dumped down the hillside and, possibly, slate was taken up the slope for loading on the P&DR above. It is possible to walk along the cutting, but it is slippery and difficult, ending in a sheer drop down into the quarry, so it requires very careful walking.

Looking to the left from the track at this point, we are overlooking the riverside path again. It is possible to see the extent of the loose spoil, now forming the largest part of the hillside and obscuring the original shape of the land. It is worth remembering that any hillside which is composed of loose pieces of slate is not natural; the whole of this hillside for about 200 yards is spoil, without which the area would look very different to that which we see today.

Plymbridge Slate Quarries

Continuing up this path as it zig-zags up the hillside, it joins the Plymouth & Dartmoor Railway line, which runs along the hillside with the modern Wrigley's factory above. The railway track, complete with granite sleepers called setts, can be followed for some way in each direction, but at this time we are interested in the quarries. Turning left along the track and walking a short distance, the remains of the un-named and long disused quarry can be seen – in fact, the track runs on an embankment right across the entrance. If we turn around and head back in the opposite direction, we come first to Higher Rumple Quarry, by far the larger, with its own siding allowing access to the P&DR, then the long, narrow Colwill Quarry, scene of the explosion which killed two people. The remains of the walls of the blacksmith's shop where the accident happened, set into the hillside to the left of the quarry entrance, can still be seen.

Carrying on along the track some two hundred yards, past the pond on the left which is in the Wrigley's factory grounds, a track on the right between

Cann Quarry from Cann viaduct.

Railway cottages now.

two granite gateposts will take us back down to the Rumple end of the Cann Viaduct. We can now cross the viaduct to the Cann side of the river, stopping to have a look at the quarry as we do so.

The first thing we see as we walk along the old railway track is the ruins of Railway Cottage on the left. An information board attempts to display what life was like here. Stand back and take a good look at this cottage, then think back to 1851, when there were twelve people, six of whom were adults, living there in the two separate halves. Lost amid the undergrowth to the left of the cottage is a building thought to have been the blacksmith's shop. A natural spring, which still runs today into a cistern further up the path, was their source of water.

Walking past the front of Railway Cottage, we find a path which takes us parallel to the railway track, up the hill past the stone bridge which straddles the line here. Continuing up the hill as it curves around to the left, we pass between a pair of walls and find ourselves looking at the ruins of a substantial building. This was Cann House, home to Thomas Pearson and James Soper, and very few others. Stand by the front door, looking down the valley towards Plymbridge, and imagine the scene in 1828 when it was new and there were no trees to obscure the view. What a magnificent place this would have been,

Cann House now.

Plymbridge Slate Quarries

built on a plateau hewn out of the hillside, gardens in front and to the side, outhouses and stables at the rear. There was no railway then, of course, so no bridge over it. The path wound down the hillside towards the orchards lining the slopes below. The quarry was out of sight behind the hill, although when they were blasting you would have had to watch out for flying rocks! Access was along one of several tracks through the woods; one, known as the Forty Foot Drive, led to the Plympton to Shaugh Prior road near Woodside Animal Sanctuary. Another joined the road from Plymbridge to Plympton; the large, well-built entry gates to this drive can still be seen today on the left where the Plymbridge road climbs out of the wood.

Retrace your steps down past Railway Cottage and continue down the path beside the viaduct leading down to the quarry itself. As you pass the side of the viaduct, note the brick-filled arch, done to improve the stability and strength of the viaduct, built as it was on loose spoil. The smaller arch beside it used to have its wooden former still in place, until some moron decided to burn it.

Ahead of you as you come to the bottom of the slope are some buildings; ahead on the right, the more substantial buildings were the office, machinery

Buildings in Cann quarry now.

shed for the waterwheel and possible a shed which housed the boiler for a steam engine. To the left of these is the waterwheel pit. There were two wheels here – on the right, an undershot one which was powered by the water of the mill leat which feeds through a tunnel from the river above the weir, then continues underground to the start of the leat proper around the corner. This was the wheel which Harry Gullett modified to generate electricity in the early twentieth century and some parts remain. The second wheel was powered by water from far up the Horrabrook, brought along a leat through the forest and carried to the much larger, 30 feet diameter overshot wheel on wooden launders. To the left of the wheel pit are the remains of the sawing and planing sheds which used to run out almost to the riverbank. The old communicating footbridge which spanned the river was near the end of this shed.

Standing opposite the bottom of the slope and facing back up, you will see the remains of a small building, a workshop of some kind, which stands at the side of a man-made ditch. This ditch gives the impression of being part of the canal or leat, but the leat has always ended around the corner, even back as far as the 1839 map and certainly before the building of the railway viaduct. This ditch appears to have been built to keep the track of the tramway as level as possible as it enters the quarry. To the left of the old workshop is a large, solidly-built square plinth which may have been the base of a crane for loading the slate onto wagons.

All in all, there is not much left to show for so much labour by so many people over such a long time, except for the enormous holes in the ground themselves.

Having studied the history of these quarries, I can't help feeling that too much was expected of them. They contained a resource which could have provided a decent living for a few hard-working people, but instead many people thought they could exploit this resource to make huge sums of money, and most ended up losing huge sums instead. The only one who appears to have had the right idea was the final operator of Cann, Edward Gullett. He worked hard and provided a living for himself and his workers, paid his rent and apparently enjoyed himself. If the quarries had been run throughout by people like him they might have been more successful, but then of course they would not have made as much potential profit (or loss) for the landowners and speculators. Pity!

Plymbridge Slate Quarries

Stone plinth in Cann quarry.

PLYMBRIDGE AND BORINGDON SILVER AND LEAD MINES

In previous chapters we have seen that there were tin, copper and iron mines in the Plym Valley. Now we find something which may seem even more surprising: lead, zinc and silver lodes. These three minerals are frequently found together, and here they were found in the valley which runs eastwards from Plymbridge, parallel to and below Plymbridge Road as it climbs the hill en route to Plympton.

The metallic ores found in Devon and Cornwall - lead, silver, zinc, arsenic, copper, tin, and iron - were formed in a liquid state, floating on liquefied igneous rock, in our case granite.

Deposits of sedimentary rock were later formed to cover the ores. Twisting and buckling of the Earth's crust over the millennia caused splits in the rock formations, which has brought the seams of ores to the surface in some areas, allowing rich lodes to be readily discovered and worked. But the same phenomenon can mean ruin for a mining company when a seam of ore suddenly comes to an abrupt end at a split, leaving the miners digging up, down and sideways trying to find where the lode continues.

In normal circumstances recovering these ores in early years was only possible where they appeared at the surface because, with no means of pumping out the water which would inevitably collect, deeper mines would simply flood. Being located in a valley, however, would have meant that it might have been possible to follow the lode underground for some distance by digging a horizontal shaft or adit, which would give an idea which way the lode ran. Initially, lead ore would have been found on or near the surface and simply dug out from shallow pits in the same way as in the earlier tin and iron mines which were described in previous chapters. These pits could here be extended underground, as long as the lode did not descend too steeply. It was only the advent of water wheels, and later steam engines, providing the large amounts of power necessary to pump the water out from deep mines, which allowed such mines to exist.

The owner of the land would also own the minerals underneath it, so the initial prospectors would have needed the permission of Baron Boringdon, later Lord

Plymbridge and Boringdon Silver and Lead Mines

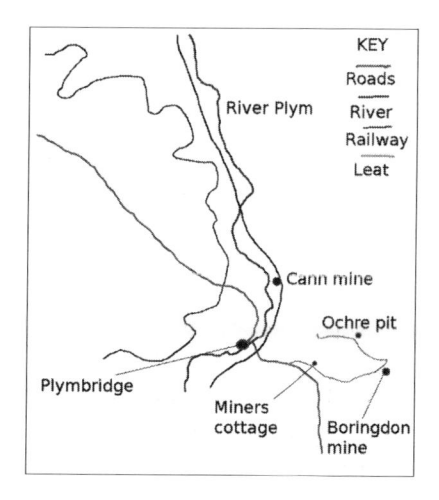

The mines around Plymbridge.

Morley, who would grant a lease – known as a sett – generally for 21 years. At first the mine would not pay its way and the landowner would charge a fixed annual rental. For a profit-making mine, the land owner could normally take 1/3 of the gross produce in place of rent, although Lord Morley was much more generous, taking only 1/10 or even 1/15 on his setts. If the mine proved to be uneconomical the lessees could give up the sett (this was known as a retiring clause in the lease), giving notice to the landowner and selling him all on-site plant at an agreed valuation.

Lead ore – the main product of the Plymbridge area mines – was often 75 per cent pure before smelting. The assay of ore, an analysis of its purity, was based on the number of parts of metal in twenty parts of ore. This gives a figure of hundredweight measures based on a ton of ore. Silver, usually found as part of lead ore, was assayed at ounces per ton. A good silver content would be 60 ounces per ton.

There were originally three mines in the Boringdon Wood area near Plymbridge (now usually known as Cann Woods). They were Boringdon Park, East Boringdon and Wheal Harriet Sophia, and all initially were worked by separate companies. In 1852 Boringdon Park and East Boringdon, which both worked the same lode, joined forces and became Boringdon Consols. Wheal Harriet Sophia was described as being on the other side of the Forty Foot Drive from Boringdon Park. The Forty Foot Drive is shown on the 1866 OS map as being the main track through the woods from Cann Quarry to what is now the Woodside Animal Sanctuary on the Plympton to Shaugh Prior road. The site of this mine has not been positively identified, although there are possible mine workings not far from the Cann Wood car park which appear to fit the description. Little is known about this mine. Worked in 1859, it was said to have been named after Harriet Sophia Coryton who had married Edmund, the 2nd Earl, in March 1842. No production figures are available.

The remains of more mining activity, which some have thought was Wheal Harriet Sophia, can be found on top of the hill above Boringdon Park mine. This is thought to have been an unsuccessful mine originally called Wheal Reynard. It was used as an Ochre pit. A lease dated 25 March 1865 between the Earl of Morley and Mr Richard Moore allows him, 'To search and work for ochre clay (iron oxide) in Boringdon Park Plantation, Plympton St Mary and Spire Pool. Term: 21 years.' Spire pool is the pool beside Plymbridge Road at the bottom of the hill.

Boringdon Park (Lead/Silver/Zinc)

The earliest mention of this mine is in 1820, when it was described as being worked by a Captain Remfry who had taken over from a previous, un-named owner who had died. The site is only ¾ mile from Boringdon Manor, which had been the home of the Parker family before they moved to Saltram and was at the time used as a farmhouse. Lead ore had been discovered at the surface, and an adit 1,200 feet long was driven on the course of the lode, but because it started in the valley and the land above is flat here the adit was never more than 72 feet below the surface. Captain Remfry's working ceased in 1824 and the equipment, which included a 23-foot water wheel, was advertised for sale in the *Royal Cornwall Gazette* of 17 July of that year.

In 1834 Captain Bray took on the sett. In 1836, a 30-inch steam engine designed by William West was ordered from Harvey's foundry in Hayle, Cornwall. The working closed again in 1839 due to the inability of this engine to pump water out from the workings, which had by then reached a depth of 120 feet.

In 1837 the following buildings were listed as being on the site:

> Counting House, Assay Office, Smith's & Carpenter's shops & iron stores, sawpit with roofing & windlass, powder house, privy, engine house, boiler house, coal yard. Total value of buildings given as £138 3shillings.

East Boringdon (Lead/Silver/Zinc)

This is really an extension of the above mine sunk on higher ground, only 230 yards away and on the same lode. The same adit was used, being extended to a total length of ¾ mile to enter the East Boringdon shaft 190 feet down. Arsenic may also have been extracted here, as the ground has a high concentration. The one shaft has recently been filled in.

BORINGDON CONSULS (LEAD/SILVER/ZINC)

In 1852 the above two mines were merged under the new title with a 40-inch cylinder pumping engine erected to pump the Boringdon Park shaft. Flat rods were used to take some of the power from this engine for pumping at the East Boringdon shaft. Flat rods were beams of wood, joined by hinges at their extremities and laid on rollers. They were used to power pumps, etc. at shafts away from the main engine shaft. There was also a 33-foot diameter water wheel, two feet wide, which provided the power for crushing the ore.

The following shafts are named for the Boringdon Park mine: Murchison or Engine shaft (240 feet deep); Hitchins (210 feet); and Annie and Morley (430 feet). The latter two could refer to the same shaft, but there are the remains of three shafts to the west of Engine shaft so my assumption is that they refer to different ones. There were workings in these shafts at 90 feet, 180 feet and 240 feet below adit level. The working at the East Boringdon shaft was 290 feet below adit level.

Between 1852 and 1857, the time of its closure, the mine produced 400 tons of lead (galena) ore, 10 tons of zinc blende (blackjack) ore, 8,000 ounces of silver and a small amount of copper. The lead and copper alone, according to the *Mining Journal*, was worth £8,000, no small sum in those days.

During this period, also according to the *Mining Journal* entries, Mr Josiah Hitchens had originally been manager, but later his place was taken by Mr James Wolferstan. The Captain throughout had been William Godden, and the secretary Mr J. H. Murchison. Hitchens and Murchison are the people after whom the shafts are presumably named.

The sale of equipment when the mine closed included the steam engine mentioned above and a 38-foot diameter wheel. According to the December 1859 *Mining Journal*:

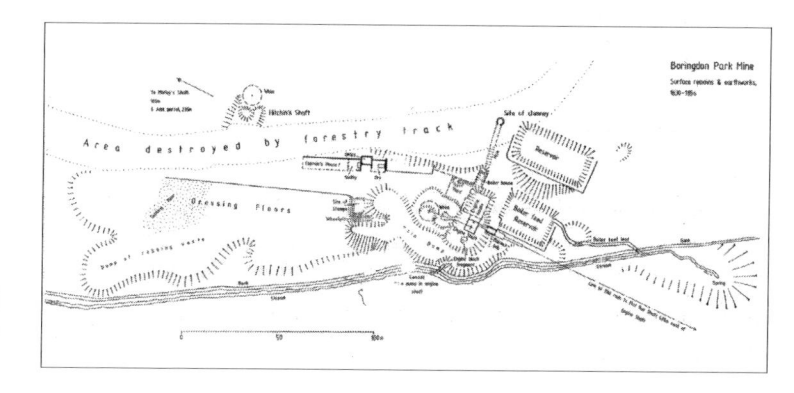

The remains of Boringdon mine.

It is generally stated that the stoppage resulted more from financial embarrassments than the bad prospects of the mine. Captain Godden entertains a very high opinion of the ground and there is a probability of its working again. The dues were 1/15th but if there is a tangible prospect of its again working a reduction of these terms will no doubt be readily obtained from so liberal a lord as the Earl of Morley.

It is possible that Captain Godden was, however, something of an optimist. Between 1851 and July of 1856 the *Plymouth & Devonport Weekly Journal* carried reports labelled 'Mining Intelligence', most, if not all, of which were attributed to W. Godden, regarding the state of work at local mines, mainly Boringdon Consols. The reports were almost always positive, including 'the lode is looking splendid, a more promising and richer lode I never saw', 'the lode is larger, and has a splendid appearance', 'the rocks of ore raised a short time ago could not be excelled at any mine' and 'leads to the confident expectation of a rich deposit of ore not far off'.

In many cases these glowing reports were accompanied by calls on the shares, i.e. requests for more money from the shareholders. The final two quotes listed above were reported on 31 July 1856. Only four months later, on 11 December 1856, the following advert appeared.

The Boringdon Consols Lead and Copper Mines.

MR HENRY WILLS, OF PLYMOUTH
Has been instructed to SELL BY AUCTION, in One Lot, on the Mine, near the Plympton Station of the South Devon Railway, Six Miles from Plymouth, on TUESDAY, the 23rd December at One O'Clock pm., all that
VALUABLE MINE CALLED BORINGDON CONSOLS,
With the LEASE, MACHINERY, MATERIALS and BUILDINGS, comprising an excellent 40-inch PUMPING ENGINE of modern construction manufactured by Harvey and Co., of Hayle, a capital Water-wheel, and crusher attached; Flat rods about ¼ of a mile in length; Tramroads & c., all in good working order. The mine is advantageous situated for the transit of materials and ore to the shipping port by rail. Upwards of £7000 worth of ores has been sold by the present company, and a considerable sum expended in the workings but it is confidently expected, by practical miners, that with a small additional outlay the mine which is now in full working order, might receive a more complete trial by a further exploration in depth and that there are also many other interesting points well worthy of prosecution.

By 30 July 1857 the following advert appeared in the *Plymouth and Devonport Journal*:

Plymbridge and Boringdon Silver and Lead Mines

Boringdon Consols mining company notice to creditors. All persons having any claim against this mine or the adventurers are hereby required within one month of the date hereof to furnish me with accounts thereof in writing, in order that the same may be investigated and if correct paid and unless such claims be made to me the assets will be appropriated and the adventurers will not hold themselves liable for any claim or demand whatsoever after the expiration of the above period of one month.

Dated 15th July 1857, J.H. Murchison, Secretary, 117 Bishopsgate Street, London

So possibly Godden was guilty of being rather optimistic!

There are many remains existing of these extensive mining operations. The area where the main workings were situated is part of Cann Woods, which is owned by the Forestry Commission and has an open-access policy. The Boringdon Park area has been extensively cleared recently, removing all the undergrowth and trees and allowing the site to be inspected much more easily. All the main shafts, plus some air shafts, have been cleared but remain cordoned off – the shafts have been capped but I would not want to be the one to investigate how well they were capped!

The area around the old East Boringdon shaft is visible in a wooded area in the middle of the privately-owned field adjoining the main Boringdon site, although there is little to see there.

Miner's Cottage, the present National Trust office and formerly the mine captain's cottage, is obviously the best evidence of the workings. Searching through the upper end of the valley above Miner's Cottage, it is possible to find the remains of two large reservoirs, one of which fed the boiler of the steam engine and the other the water-wheel. Various walls remain from the engine house and associated buildings, plus those from the smithy, office and a building described in the map above as a dry, which contains plates from what looks like another boiler. There are dressing floors, where the ores were broken down prior to refining, plus extensive waste tips.

The shafts themselves have been filled in, but it is possible to find their sites. Engine Shaft was, inevitably, beside the engine house. Hutchins Shaft is on the hillside above, and further down towards Miner's Cottage is the remains of another shaft on top of the hill, which was called Annie's or Morley's Shaft. Near to Miner's Cottage, hidden now behind a shed, the portal of the main adit was found by the then National Trust warden, Stephen Holley, who supplied me with much of this information and the map above. A further adit is easily visible because it opens onto the bank of the Cann canal, some hundred yards north of where the canal leaves the Plymbridge Road. Although I have no proof, this must logically be associated with this mine because the nearer Canal mine is several hundred yards further north on the other side of a valley.

Above left: The remains of the boiler.

Above right: A canalside adit.

Boringdon Tin Stream Works, 1836–48

Little is known for certain of the situation of this mine except that the works were described as being in a marsh on the east side of and adjoining the River Plym. The *Plymouth and Devonport Weekly Journal* had two reports, in July and September 1836, about the opening of the venture, describing it as 'Plym marshes' at Plym Bridge, and being in a 'beautiful vale on the Plym river', with 'easy access by tram road'. Logically this would point to the marshy area running from Plymbridge Lodge up the valley beside Plymbridge Road, to the east of the river. The Cann Quarry tramway, which ran across the entrance to the valley, had been opened the year before.

After the above reports, we have little information of these works until an advert in the *Plymouth, Devonport and Stonehouse Herald* of 28 July 1849 stated:

> Between Longbridge and Cann slate Quarry, about 4 miles from Plymouth, new mining materials and machinery to be sold at auction by George Trickett on Tuesday, August 7th, at 12 O'Clock at the Boringdon Tin Stream Works. All the excellent material and machinery in and upon the workings consisting of one waterwheel 18 foot diameter and 6 foot breadth with wrought iron cranks, quite new, brasses and bearings one old waterwheel, flush hatches, launders props and stay, pendulum bobs, cast iron bobs, 1¼ inch flat rods, flat rod pulleys, triangles, hand pumps, two 10 inch working barrels, one 10 inch door piece and door, 2 ton inch wind bores, one 3 foot matching piece, 6 foot 11 inch plain pump, one 4 foot 11 inch plain pump, one 9 foot 11 inch plain pump, one cast iron matching piece, 2 loggerheads and brasses, old iron, quantity of useful timber, and sundry other useful materials.

Plymbridge and Boringdon Silver and Lead Mines

The Cann Quarry railway running alongside the Mine affords a cheap means for the carriage of the above.

This seems to be the end of this venture, and once again this would point to the marshes near Plymbridge Lodge as the site. The 1840 Tithe map shows nothing there or anywhere else nearby, not even Cann and Boringdon Mines, which were definitely there, so we have to presume mines were not shown on tithe maps. There was a later operation in this area, the Spire Pool near Plymbridge Lodge, which is mentioned with regard to the search for ochre in 1865 and there is also thought to have been an attempt to rework the Boringdon Consols spoil tips to recover more metals, or possibly arsenic, from them. This is shown on the 1886 OS map with settling tanks and a chimney, presumably for a calciner, a type of furnace in which the ore was heated to allow better extraction of the minerals and also to allow arsenic 'soot' to be recovered from the flues. The chimney and the walls of the settling tanks still exist below Miner's Cottage.

CANN MINE (LEAD/SILVER)

This mine, also known as Canal Mine, is roughly two-thirds of the way between Plym Bridge and Cann Quarry, and is located between the canal and the old railway track, now a cycle track, above. All of this area is National Trust property. The lode

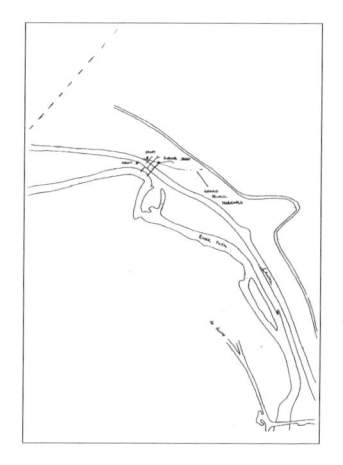

Above left: A tracing of an 1829 map of Cann mine.

Above right: A view down the shaft of Cann mine today.

was discovered in 1824 when excavations were being carried out to form the Cann Canal from the original mill leat, as described elsewhere, in order to take the slate out of Cann Quarry. It was worked between 1824 and 1825 by Petherick of St Blazey, Cornwall.

The sett was re-granted in 1829 to one Joseph Thomas Austen; he had been born in Plymouth in 1782 but changed his name to Joseph Austen Treffry in 1808 when he inherited his uncle's estate in Fowey. He went on to be a major landowner and industrialist in Cornwall, developing Par harbour and opening several mines in the area. Sadly, Cann mine was not one of his success stories! An above ground survey from 1829 showed three shafts – one between the river and the canal and two on the east of the canal. According to Bert Shorten, in his book *Plympton's Old Metal Mines*, the mine used a 'small' steam engine and a 22-foot water wheel to power the pumping, lifting and crushing operations.

The mine was inspected in 1832 by Charles Thomas, captain of Tolcoath Mine. He reported that it lay in a slate formation that contained 'a little copper', which was not exactly a glowing recommendation for the mine!

Today there is an open mine shaft with two side workings visible, the base of a demolished chimney, the remains of a wheel pit, the ruins of what was probably an engine house, other ruins which could have been the site of ore crushing/stamping machinery and a possible ore-dressing floor. There is also evidence of spoil heaps, and a possible second filled-in shaft but I can find no sign of the shaft supposed to have been between the canal and the river. The open shaft is accessible with care from the cycle track above and has a surrounding fence and a large steel grating over it, but it is just possible to see down the shaft and inspect the entrances to the side workings in relative safety. No production figures are available for this mine.

MARSH MILL

The Marsh Mill was an ancient mill and must have been a very successful one, simply because it lasted so long. It was initially known as Woodford Mills because it was part of the Manor of Woodford. According to Brian Moseley's excellent 'Plymouth Data' website, in 1582 one Jerome Mayhowe was granted the right to take stones, turf and timber from Boringdon woods in order to keep the leat and Woodford Mill in good repair. The Mayhowe family were at that time the owners of Saltram and large sections of land in the area, and it was in that year that the Parker family inherited Boringdon House.

The next documentary evidence relating to the mill, or at least the leat for the mill, comes in a lease of 1723 for Cann Quarry which includes the following: 'Occupiers to supply stone and workmen where required for repair of bartons or farms, also stone for the repair of the head-weir, salmon hutch and leat.' The head-weir was there to provide sufficient head (or depth) of water to ensure the flow to the leat, and as can be seen in the advert below this seemed to work very well. From 1829 onwards the leat had been widened to form the Cann Quarry canal, so from then on there would never have been a shortage of water!

The mill had two water wheels and four pairs of stones for grinding grain (i.e. a grist mill) according to various adverts in the eighteenth and nineteenth centuries. It also had two bolting, or sifting, machines so that the ground flour could be graded by sifting out chaff and coarser lumps.

The origin of the name Marsh Mill has sometimes been the cause of some argument. The obvious reason for the name is that, although built on higher, drier ground between the River Plym and the Tory Brook, it was certainly just across the River Plym from an area of marsh land known as May's Marsh, which was in the area where the river was tidal and would hence have been very prone to flooding at high tides. This area is now covered by part of Marsh Mills Retail Park and the Riverside Caravan Park.

An alternative source for the name is the fact that from 1795 until at least 1826 it was also operated by Thomas Marsh of Bridport and Henry Marsh of Plympton St Mary, who despite their shared surname were unrelated. Unfortunately for those

Industrial Archaeology of the Plym Valley

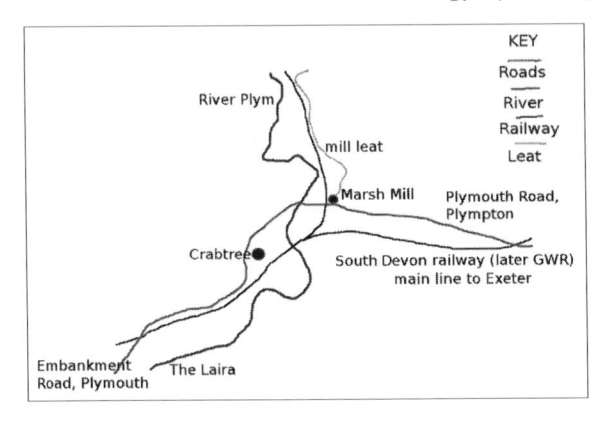

The location of Marsh Mill.

favouring this argument, the following advert was placed in the *Sherborne and Yeovil Mercury*, dated 6 September 1784:

> To be let for a term of 7 or 14 years from Michaelmas next, all that commodious set of water grist mills with a good and convenient dwelling house, hoggery, stable, and other outhouses, commonly called or known by the name of Woodford or Marsh Mills together with 3 closes of exceeding rich land and an herb garden adjoining thereto, containing in the whole about 12 acres, situate, lying and being in the parish of Plympton St Mary, in the County of Devon.
>
> The said mills are distant about 3 miles from Plymouth, adjoining the turnpike road leading from thence to Exeter, and about a furlong from a navigable part of the river Plym. They consist of 4 pair of stones, 2 bolting mills, a proper hoisting tackle, and have been entirely rebuilt in the course of 4 years. They were never known to want water in the driest Summer, or to be impeded by the frost in the severest winter, and are in every respect well calculated for carrying an a very extensive flour trade.

This would appear to show that the name was already being changed from Woodford to Marsh Mill before the Marshes took it over. John Rider of Torbryan, miller, was the person who took advantage of that particular advert, but Messrs Marsh ran the mill from 1795, and at least one of them was still involved with the mill on 26 October 1826 when the following advert ran in the *Plymouth and Devonport Weekly Journal*:

> Capital flour mills to be let for a term of 7 years from Lady Day 1827, all those capital flour mills called Woodford or Marsh Mills, situate lying and being in parish of Plympton St Mary, in the county of Devon, in the immediate neighbourhood of

Marsh Mill

the populous and increasing towns of Plymouth, Stonehouse and Devonport, close adjoining the Plymouth eastern turnpike from which town it is only about 3 miles. The mills have four pair of stones, two waterwheels, flour smut and bolting machines, have a never failing supply of water and are capable of making from 130-200 sacks of flour weekly, two lofts in which may be deposited upwards of 690 quarters of corn, a bakehouse with two ovens, together with a convenient dwelling house, stable, cowhouse, hog sties, court hutches, gardens and about 12 acres of exceeding rich grazing land immediately adjoining in Plympton St Mary aforesaid and Egg Buckland. Are now and have been for many years past the occupation of Messrs Marsh and Balkwill. The conditions on which the same will let may be seen on application to Mr Yolland at Merafield near Plympton and by whom sealed tenders received titled Friday 10th day of November, immediately after which the person whose tender is accepted will have notice. And in the meantime, the mills and other premises may be viewed by applying to Mr Balkwill thereon.

Reading these adverts, it is obvious that this was a substantial place, both as working premises and a dwelling house. A further advert appeared on 27 February 1829, when a new seven-year lease was advertised. In addition to the flour and grist mills, there were 18 acres of 'exceeding rich pasture land'. Furthermore, 'The situation and capacity of those Mills, so immediately contiguous to Plymouth, the certainty of a never failing supply of water, the improvements lately effected by the new Canal from Cann Quarry, are too well known to require any description.' This latter refers to the rebuilding of the mill leat to carry slate from the quarry. By 1835 this had been replaced by a branch of the Plymouth & Dartmoor Railway which ran most of the way along the towpath of the canal, veering off to pass some few hundred yards to the west of the mill with another, short-lived, branch to Plympton passing along the southern boundary as can be seen in the Title map.

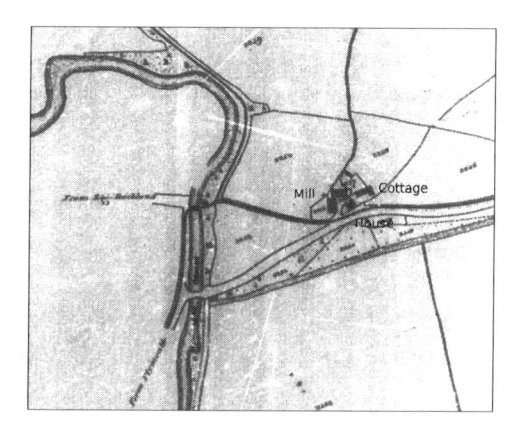

A tithe map showing Marsh Mill.

The 1840 Plympton St Mary Tithe Map gives us the first impression of what the mill looked like at the time. The mill itself appears to be made up of two substantial buildings in the form of an 'L', with the house between the mill and the road. To the east of the house is another, smaller, 'L'-shaped building which is presumably the cottage.

By the time of the 1841 census Richard Shilson, 45, was the miller, with his wife and family. Thomas Gullett, 25, worked for them as a mill servant, Jane Hall, 20, and Eliza Gullett, 16, were family servants.

Incidentally, at Marsh House, across the river from the mill at Crabtree, was the large Daw family, also millers and of whom more anon; George, 45, his wife Jemima, 43, their five daughters and five sons and three servants.

At this time the mill was not being operated solely as a grist mill but was shared between Richard Shilson, the grist miller, who had two thirds of the mill and the stream, and William Alsopp of Plymouth, an earthenware manufacturer who had use of the western wheel and one third of the mill and stream. According to this complex agreement, the use of the stream was on a time basis, with Shilson having occupation from 6 a.m. to 10 p.m. and Alsopp from 10 p.m. to 6 a.m. Alsopp traded as the Plymouth Pottery Co. When he left in 1858, he took with him two flint mills, a glaze mill and a little colour mill.

From 1845 the operators of the mill were listed as George Frean and George Daw who, as we have seen, had been living at Marsh House. At around this time, Frean was also operating Drake's Mill, the site of which is the gardens next to Drake's Place Reservoir in the centre of Plymouth. His son, also George, later went to London to manage his father-in-law's new biscuit factory, from which grew the famous Peek Frean's.

The 1851 census shows that Richard Lewis, 33, was the miller with his wife Eliza, 32, and their family. Thomas Gullett is still there, now aged 50, his Irish-born wife Mary, 43, their sons George, 19, Thomas, 13, and William, 10. The first two sons are also millers, as are the lodgers, James Renshols and John Hambly. Although there is no indication of how many in total were employed there at that time, business must have been good to have had so many millers in residence.

In 1859 the Tavistock & South Devon Railway opened, passing down the western edge of the property; the mill was even provided with its own siding for ease of loading. With the main Plymouth to London road passing on the south side, it could hardly have been better furnished with transport connections.

By 1861 George Daw was himself listed as the miller there. Aged 55, he described himself as 'Corn factor and miller employing 11 men and 1 boy'. Their daughter Rhoda Frean, 24, is a 'merchant's wife', having presumably

Marsh Mill

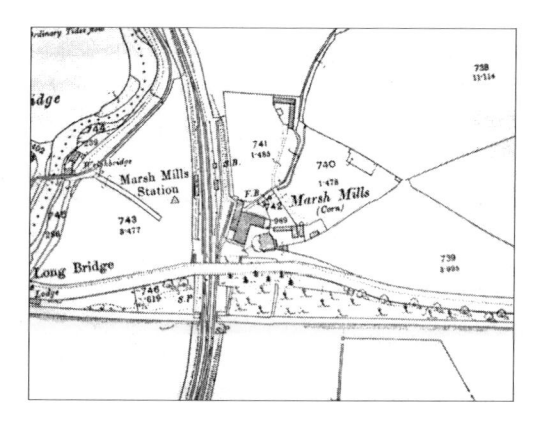

An 1894 map showing Marsh Mill.

married a member of the family of Daw's business partner.

For the first time, Marsh Mills Cottage (the smaller 'L'-shaped building shown on the Tithe Map) is listed separately; previously, there were simply multiple listings under the heading 'Marsh Mills'. Living at the cottage were James Lewis from Chagford, 52, a journeyman miller, plus his family. Also living there were Thomas Gullett, 59, now a waggoner, and his family, of whom one son, William, 20, was a journeyman miller. Lodging with them was James Kenshole, 22, also a journeyman miller. A journeyman was a tradesman who was hired, in many ways like a self-employed tradesman today.

It can be seen from the above list of inhabitants that the mill was by this time a very large concern, employing '11 men and a boy', with at least three millers, apart from the proprietor, and their large families all living on the premises. The 1894 map shows that it had grown considerably in the 54 years since the tithe map.

A decade later Richard Daw, 28, second son of George, was the miller with his family and servants. Marsh Mills Cottage had John Williams, 41, a miller, and his family.

By 1881 the miller was still Richard Henry Daw, 'miller and corn merchant employing 18 men'. With an increased workforce, the mill must have been prospering.

The census also shows, presumably at the cottage, George Ash, 40, a miller's waggoner, plus his family. Also there were James Rowe Scoble, 35, a miller's labourer, with his family plus a lodger, George Francis Congleton, 25, a miller.

Richard Henry Daw was still there in 1891, although he was no longer boasting of how many employees he had. There were no other millers listed at the premises – George Ash is still the waggoner.

By 1901 Richard and Louisa Daw were still there; he was then 58 and described as a 'retired corn miller'. At the cottage were James Guest, 32 and a

Industrial Archaeology of the Plym Valley

Marsh Mills letterhead, 1906.

Above left: The Marsh Mill, 1914. (*John Boulden*)

Above right: A document giving information about the auction of Marsh Mills.

'driver for a miller', perhaps showing the earliest signs of mechanised transport! The fact that there were no millers living there might appear to show that the mill was unproductive at this time, although this is unlikely given how successful it was.

From at least 1906 the mill was operated by Boswarva & Harris Ltd and driven now by both steam and water. The land tax listing of 1910 tells us that the mill was still owned by the Earl of Morley and leased to a company

Marsh Mill

called Marsh Mill Milling Co., which was the trading name of Boswarva & Harris.

The photograph of the mill in 1914 shows what a huge and seemingly successful place it was, and suddenly the boast of '18 men' employed seems very reasonable. It was still basically the large 'L'-shaped building shown on the Tithe Map but it had grown much larger and had some additions and improvements. Boswarva & Harris appear to have bought the mill from Lord Morley in 1920 and were still working in 1927. Surprisingly for such an apparently large and successful place, the firm went into voluntary liquidation and the mill was put up for auction by the liquidator on 31 October 1929. The auction reached £2,000, which was below the reserve price and it was withdrawn. The auction brochure describes the buildings throughout as 'substantial' and 'well-built', including nearly four acres of grounds with paddocks, gardens and piggeries. There is no mention in the brochure of the Cottage, which had been listed on the census as occupied as recently as 1901. This building was shown on the Tithe Map and was still there on the 1894 OS map but had disappeared by 1913.

The mill building appears to have been largely demolished after the auction. The 1933 map shows the outline walls but no roof, but the house continued to be inhabited. The tax listing for 1938 tells us that the site was then owned by Hoare Bros, who were operating a tarmac and roadstone business there from 1937 to at least 1955, the house being occupied in 1938 by one S. R. Churchill. Hoare Brothers were a large concern, big enough to have their own railway wagons, a model of which is still available today for OO gauge railway enthusiasts!

John Luscombe, who lived nearby, also recalls a roadstone crushing works there in the 1940s which was incredibly noisy but stopped work at noon, so everyone locally knew what time it was. Local directories also show several builders' merchants operating from the Marsh Mill site in the early 1950s. By 1962 Hoare Brothers were still operating but had by then moved from the site to various quarries around the area.

From the early 1940s the mill house became a dairy run by a Mrs Turpin, and John Luscombe remembers being sent there for milk by his mother. Turpin's dairy remained in operation there until at least 1981, when presumably under threat of demolition it moved to Glenholt, where it was operating in 1984. Given that the first mention of Woodford Mill was in 1582, the place had by then been the site of commercial operations for four hundred years!

Sited in a large concrete building at the rear of the house was a commercial vehicle repair garage, Leyland Motors, which operated from around the mid-

Industrial Archaeology of the Plym Valley

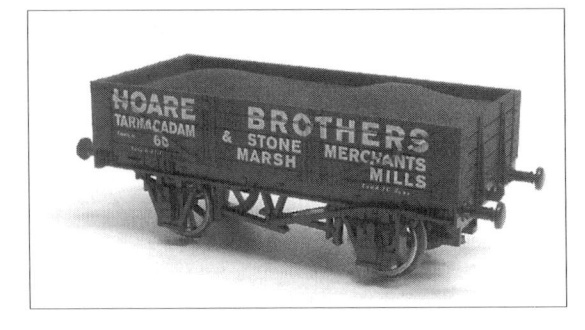

A model wagon in the livery of Hoare Bros.

Marsh Mill House from Plymouth Road. (*John Luscombe*)

Turpin's Dairy, the old mill house. (*John Luscombe*)

The rear of Marsh Mill House. (*John Luscombe*)

The ruins of the mill. (*John Luscombe*)

1960s to the early 1970s and is remembered by many local people. This was on the site of the 'spacious yard' fronting the main road which was listed in the auction particulars.

The mill was bordered on the south by Plymouth Road and on the west by the railway, so the site lies near the south-western end of the B&Q car park, close to the bridge which carries Plymouth Road past there. All the remaining buildings were demolished in the mid-1980s when preparations were being made to build the B&Q store, which opened on 19 August 1987. The last vestige of the Marsh Mills station and its railway line, now disused, which ran down the western side of the mill, can be seen between the two access roads.

EPILOGUE

Beyond the site of the old Marsh Mill, now the B&Q store, and the more modern Marsh Mills roundabout nearby, the Plym Valley becomes the Plym Estuary – the Laira – and hence outside of my remit. I hope you have enjoyed our journey down the valley and have not been too bored along the way.

While researching this book, I learned a great deal about the industries which were carried on in this valley as well as a little about the lives of the people who worked in those industries. I was amazed to find that some of the industries had existed for centuries, providing employment for often unnamed workers with which to support their families. I have also learned a little about how hard their lives were and how much easier our lives are today in comparison.

I've learned a great deal of respect for all the people who put a huge amount of energy, hard work and, in some cases, money into the various enterprises which sprang up, grew and died along the valley. Life changes little over the years; now as then, we see businesses start up with what appears to be a very good idea and the owner and his workforce do everything they can to make it work. Some succeed, some don't. Some barely get under way before they fail, others provide a good living for both owner and workers for many years, some go on for so long they just seem to have always been there. Some fail because the basic idea was unsound, others which were sound ideas failed because the owners or managers had no idea how to run them. I'm also convinced that, then as now, a few charlatans made a great deal of money by duping others, spending other people's money and getting out just ahead of the bailiffs; some got their just deserts, others got away with it.

Another thing I've learned is just how quickly nature reclaims her own. Places which just fifty years ago were inhabited, or were busy working sites employing many men, are now overgrown with few signs of anyone ever having been there. Trying to photograph these places has brought this home to me, mainly because there are now trees everywhere. A viewpoint which once provided an uninterrupted vista now shows nothing but trees! For instance,

Epilogue

standing at the viewpoint from which the 1831 etching of Cann House was done, it is impossible to see Cann House at all because of trees on both sides of the valley. Similarly, the wonderful view of the Brunel timber viaduct at Cann, taken from atop the spoil tip at Rumple, is impossible today. Possibly the best example is the photograph taken at Shaugh Bridge platform early in the twentieth century, just over 100 years ago, from which at the time it was possible to see the railway sweeping off through open country with just a few trees, all the way to the Leebeer Tunnel entrance a quarter of a mile away. Today, the area is thick woodland and you are lucky to see anything much beyond the end of the platform! A couple of years ago I was involved in helping to clear the Shaugh Lake clay kiln of some of its trees and bushes; it has been cleared again since then and will continue to need such work doing regularly, otherwise it will become totally lost in the undergrowth, like Riverford is now.

If as a result of my writing this book you have been encouraged to explore an area you have never previously visited or you have discovered a new place to walk, that is good. Maybe you already walk there and for years have thought, 'I wonder what that place was?' If I have provided the answer, then that is even better.

Few cities of the size of Plymouth are fortunate enough to have such a beautiful wooded valley so near to the city. Even allowing for the city traffic, from my home in central Plymouth I can be in the nearer parts of the valley within 15 minutes, and the furthest parts within 30 minutes at most. A hundred years ago much of it would have been private land or even an industrial site, with little or no public access. Even fifty years ago parts of it were not open to the public, so we are very lucky to have it so close by, and long may it continue to be somewhere to walk, enjoy the beautiful surroundings and see the abundant wildlife. In our crowded, high-speed modern world we need such places more than ever.

ALSO AVAILABLE FROM AMBERLEY PUBLISHING

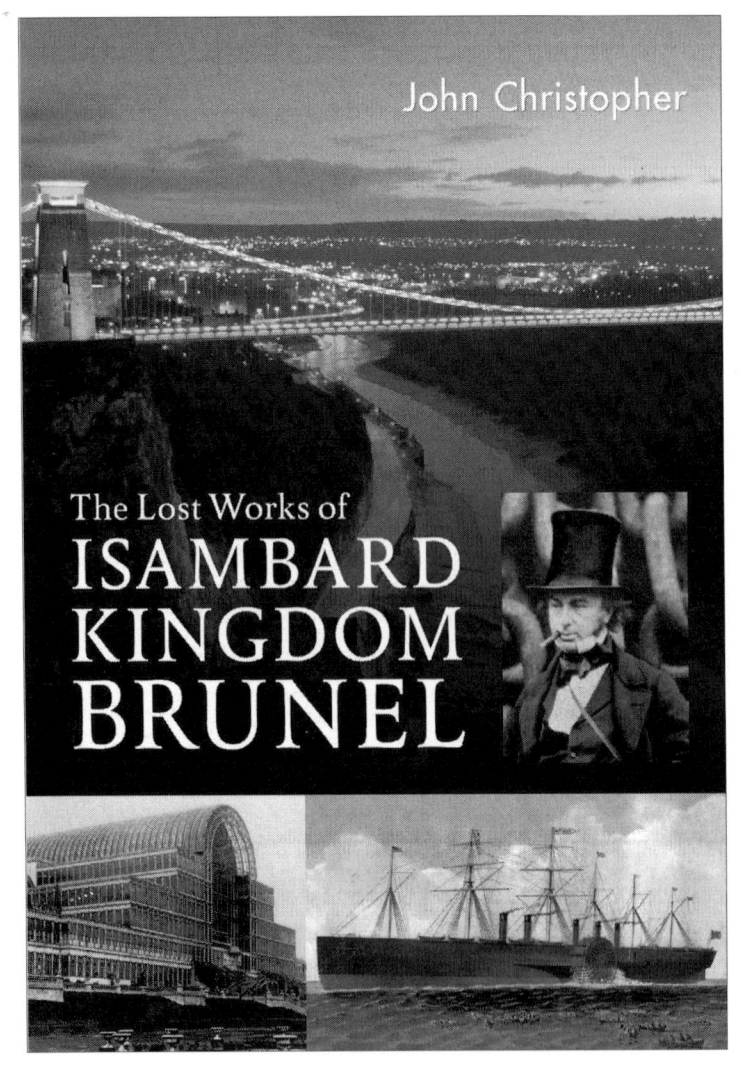

The Lost Works of Isambard Kingdom Brunel
John Christopher

The first history of Brunel's lost works, by an acknowledged Brunel expert.

978 1 4456 0090 1
96 pages, full colour

Available from all good bookshops or order direct from our website www.amberleybooks.com